THE GEOMETRY OF SHEET METAL WORK

FOR STUDENTS AND CRAFTSMEN

BY

A. DICKASON

C.G.I.A., A.C.T. (BIRM.)

PITMAN

PITMAN PUBLISHING LIMITED
39 Parker Street, London WC2B 5PB

Associated Companies
Copp Clark Ltd, Toronto · Fearon Publishers Inc, Belmont, California
Pitman Publishing New Zealand Ltd, Wellington · Pitman Publishing
Pty Ltd, Melbourne · Sir Isaac Pitman Ltd, Nairobi

Reprinted 1967, 1969, 1970, 1972, 1974, 1976, 1977, 1978

Reproduced and printed by photolithography and bound
in Great Britain at The Pitman Press, Bath

ISBN 0 273 41513 1

THE GEOMETRY OF
SHEET METAL WORK

PREFACE

IT is a fairly obvious fact that skill in the manipulation of the hammer can be an effective time-saver, but it is not so readily seen that skill in the application of the rule and compasses can be a greater, and even more effective, means of economy. Pattern-drafting cannot be dispensed with. Whether it be the preparation of templets to be used in mass production, or the drafting of a single pattern to be used once only, marking out has to be done, and this can be executed with speed and certainty by craftsmen well versed in methods of geometry. Mental haziness in this connection is largely responsible for errors and undue delay.

Versatility in the art of pattern-drafting is a valuable asset to the sheet metal worker, and the ability to adapt one's knowledge of surface developments to fresh problems depends on a clear grasp of the underlying principles rather than on a superficial learning of haphazard collections of problems. If the study of pattern-developing is to be profitable, it should be founded on principles which can be systematically arranged to form a progressive plan or course. Examples should be classified so that they develop a line of thought or illustrate a fundamental principle.

An important feature of this work is the classification of method as applied to the development of surfaces. The methods of Radial Lines, Parallel Lines, and Triangulation are emphasized by the arrangement of the problems and the grouping of similarities in methods of treatment. This arrangement simplifies the study of the principles used in pattern-drafting, and since all patterns may be developed by one or other of the three methods mentioned above, a clear understanding of the principles involved will assist the student or craftsman considerably in gaining reliability as a pattern-drafter. The work is divided into courses as a further help in the grouping of ideas and principles. The first three courses deal progressively with the three methods of development. In the Fourth Course the method of Cutting Planes applied to the solution of problems of intersection is discussed at some length. In the Fifth Course, the methods of Double Projection find application in the solution of difficult problems of pipe intersection. Twisted surfaces and spiral chutes are also included in this course.

This work first appeared in *Sheet Metal Industries* in the form of articles, and I should like to take this opportunity of expressing my thanks to Mr. A. McLeod, Editor of *Sheet Metal Industries*, for the valuable assistance he afforded me in getting this work completed. My thanks are due, also, to many friends and colleagues who have shown a keen interest in my work, and helped in no small measure in the development of my plans.

A. DICKASON

CONTENTS

FIFTH COURSE

CHAPTER 15

CHAPTER 16

THE GEOMETRY OF
SHEET METAL WORK

CHAPTER 1

CLASSIFICATION OF GEOMETRICAL FORMS

IF the study of pattern-developing is to be profitable, it should be founded on principles which can be systematically arranged to form a progressive plan or course. Examples should be classified so that they develop a line of thought or illustrate a fundamental principle. Those illustrious examples which introduce short cuts without due respect to a place in a formulated plan should be regarded with caution. Many of the standard problems have been so well pruned that they form examples peculiar to themselves. Such examples, presented in as few lines as possible, tend to confuse rather than to elucidate a principle. There is, of course, something to be said for the expression of a problem in its briefest form, since the saving of time is always an important consideration, but, for purposes of study, methods of abbreviation are best left alone, or reserved for special attention after the fundamentals have been thoroughly digested.

THE OBJECT OF SURFACE DEVELOPMENT

Most sheet metal articles are of geometrical form, either simple, compound, or complex. Included among simple forms are the cube, the cylinder, and the cone. Among compound forms, in which two or more simple forms are combined, might be included the watering can and the funnel. The watering can, for instance, has a cylindrical body and a conical spout. A complex form may be neither simple nor compound, but may, in itself, be of some shape not easily defined ; a mixture of form, such as a transforming piece from a rectangle to an ellipse.

Since nearly all sheet metal articles are hollow bodies, the metal sheet might be regarded as constituting the "surface." This surface is rolled or worked into form from the flat sheet, and to ensure that the finished form shall be correct to given dimensions, the pattern cut from the flat sheet should be true in shape and size at the outset.

A good deal of time and trouble may be saved in the working up if care is taken with the setting out of the pattern. The process of determining the shape of the pattern is called "developing the surface."

CLASSIFICATION OF GEOMETRICAL FORMS

The great variety of geometrical forms might be classified in numerous ways. The classification should depend primarily on the object in view. For the purpose of surface development the order of classifying should be in accordance with the methods adopted for development.

CLASS 1. If a piece of cord be suspended from a fixed point, and the other end, carrying a weight, be free to move, or swing, the weighted end can be made to describe an infinite variety of forms, such as a square, a circle, or an oval. If these forms be regarded as lying in a flat plane, instead of a spherical, then the cord, kept taut by the weight, traverses the surface of a solid having the flat shape at its base.

Since the cord is suspended from a point, it will be clear that, whatever shape the base might take, the top of the solid is a single point, or apex. This class of solid is termed a "pyramid." The cone might therefore be regarded as a pyramid with a circular base.

CLASS 2. If, now, the point from which the cord was suspended, instead of being fixed, were allowed to move and describe forms in a horizontal plane, the cord, with its weighted end would, by maintaining a vertical position, traverse the same forms at the bottom as at the top, and therefore throughout its length.

Solids of this kind have a uniform cross-section in shape and size, and are termed "prisms." The cylinder, which has a circular cross-section at any point in its length, may be regarded as a prism with circular ends.

CLASS 3. One other possibility remains. Each end of the taut cord might be made to trace out, at the same time, different patterns, such as a rectangle and a circle, a square and an ellipse, or any other combination of forms.

This type of body may be called a "transformer," since its shape at one end transforms to a different one at the other. Although both ends are different in shape, the piece of cord may be made to lie along the surface of the transformer from end to end. This is a point of great importance in connection with developments.

CLASS 4. There are other geometrical forms which have a surface curving both ways, such as the sphere, the spheroid and their frustums. In these cases the piece of cord, while being kept taut, cannot be made to lie along the surface; it can only be made to touch at one point.

METHODS OF SURFACE DEVELOPMENT

There are three distinct methods in general use by means of which the surfaces of geometrical solids may be developed. In deciding which of these three methods should be used, the class of solid to which the object belongs must be taken into account. The foregoing classification of geometrical forms should be of use in deciding on the method to be applied.

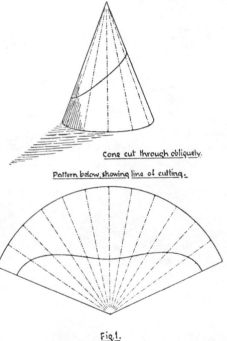

Cone cut through obliquely.

Pattern below, showing line of cutting.

Fig. 1.

METHOD 1. Take a solid belonging to Class 1—the cone, for instance—and imagine a series of straight lines to be drawn on its surface from its apex to its base. If the surface of this solid could now be unrolled like a piece of paper, and laid out flat, it would be found that all the lines drawn on the surface radiated from one point, which was the apex of the cone.

Now imagine the cone to be cut across its middle at an angle with the base. If now the surface could be unrolled as before, the line of cutting would be clearly defined on the radial lines. (See Fig. 1.)

It should be clear from this that the surface of all solids having an apex could be developed by a system based on radial lines. This includes all frustums and bodies which form parts of pyramids or cones.

METHOD 2. Take a solid belonging to Class 2—the square prism, for example. If this were rolled over on its sides one after another, or, in other words, if its sides were opened out flat, the result would be a rectangle divided into four smaller rectangles by parallel lines corresponding to the edges of the prism. Again, imagine the prism to be cut through obliquely, and that the sides were rolled out as before. The line of cutting could easily be traced by the intersection on the parallel lines.

Take now a cylinder. If a series of parallel lines were drawn on its surface from end to end, and the cross-section of any shape taken,

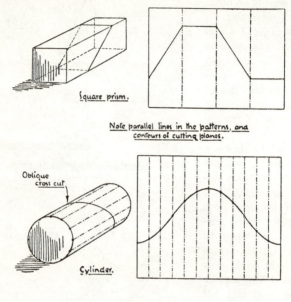

Square prism.

Note parallel lines in the patterns, and contours of cutting planes.

Oblique cross cut

Cylinder.

Fig. 2.

the contour of that cross-section could be shown in the flat by means of the intersecting points on the parallel lines. (See Fig. 2.)

It follows from this that the surfaces of all solids belonging to the class of prisms may be developed by a system based on parallel lines.

METHOD 3. Take a solid belonging to Class 3—for example, one with an oval base and a circular top. (See Fig. 3.) The sides of this solid produced do not converge to an apex, although at first sight it might appear that they do so, but a glance at a front elevation in conjunction with a side elevation will dispel any illusion on that point. It cannot, therefore, be developed by the method of radial lines. Again, parallel lines cannot, in any direction, be drawn on its surface. Therefore, it is impossible to develop it by the method of parallel lines.

If, now, a series of triangles were drawn on its surface, and the exact size of each determined successively, and laid side by side in

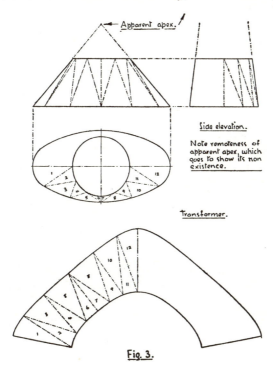

Side elevation.

Note remoteness of apparent apex, which goes to show its non existence.

Transformer.

Fig. 3.

the proper order, the complete pattern could thus be pieced together. The surfaces of all solids belonging to Class 3, transformers, admit of being divided up in this way. It follows, then, that they may be developed by this method of triangulation.

The three methods of development arising out of this classification are, therefore: (1) the Radial Line method; (2) the Parallel Line method; and (3) Triangulation. Although the method of radial lines and of parallel lines cannot be applied to this class of solids, the method of triangulation may be applied to those of Class 1 and Class 2. It is advisable, however, to use the method best adapted to the form under consideration. Solids which come under Class 4 cannot be developed accurately because the surface curves both ways and the metal has to be hollowed or raised in order to acquire the desired shape. There are methods of obtaining approximate developments, however, which more or less depend upon the nature of the material and the work to be done. These need special consideration.

FIRST COURSE

CHAPTER 2

THE RADIAL LINE METHOD

How simple it is to get into difficulties with problems of the cone! They present many pitfalls which can only be avoided by careful discrimination. For instance, the legs of the breeches piece shown at Fig. 4 (a) might be mistaken for portions of oblique cones, but an

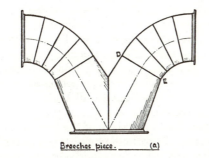

Breeches piece.   (a)

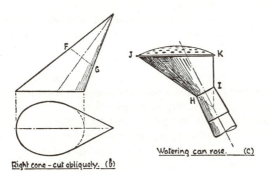

Right cone - cut obliquely.   (b)

Watering can rose.   (c)

Fig. 4.

oblique cone would present an ellipse at *DE* instead of the circle required.

Again, a body with an elliptical base, and which tapers obliquely to an apex, may not readily be identified as a right cone. Yet, when a

7

cross-section, taken at right angles to the centreline, is known to be a circle, as at *FG*, Fig. 4 (*b*), then the object is assuredly a right cone, cut obliquely. A practical example of this problem is a rose for a watering can, of the type shown at Fig. 4 (*c*). The cross-section at *HI*, at right angles to the centreline, is circular, while the shape at the top, *JK*, is an ellipse.

It might be well at this point to make a distinction between the right cone and the oblique cone.

A right cone is a body which has a circular base, and tapers uniformly from the base to a point or apex which lies perpendicularly over the centre-point of the base. It follows from this that any cross-section parallel to the plane of the base, or at right angles to the centreline, is a circle. Conversely, any cross-section other than at right angles to the centreline presents an ellipse.

An oblique cone is a body which has a circular base, and tapers to a point or apex, but the apex does *not* lie perpendicularly over the centre-point of the base. Hence, the oblique cone leans to one side. It may lean but slightly, or it may lean considerably, but the amount of its leaning does not alter its properties as an oblique cone. Arising out of this, any cross-section parallel to the plane of the base is a circle, but any cross-section at right angles to the centre line, and therefore not parallel to the plane of the base, is an ellipse.

These distinctions are important in assisting to discriminate between conical bodies and other tapering bodies such as commonly occur in sheet metal work.

The First Course of surface developments will deal with those problems which involve first principles—not necessarily the simplest, but those which naturally take first place in a combined series. The course will include—

(*a*) Developments by the Radial Line method of patterns involving problems of the right cone and its frustums, right conic sections or right cone cut off in any plane.

(*b*) Developments by the Parallel Line method of patterns for tees, bends, and elbows, and intersections of pipes of equal diameters or equal oval cross-sections ; the oblique cylinder ; elementary examples of moulding such as curbs and spouting.

(*c*) Developments by the method of Triangulation, comprising patterns of transformers for change of section of various kinds between two parallel planes, such as tallboys, hoppers, and hoods, with either perpendicular or oblique axes. The condition that the transformers lie between two parallel planes is an important one as this simplifies the study of the method.

DEVELOPMENT BY RADIAL LINES

The sides, or the sides produced, of any object belonging to the class of pyramids, and the cone in particular, must converge to an

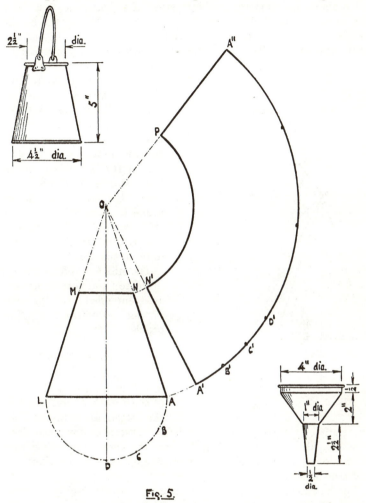

Fig. 5.

apex. Given that condition, it is possible to define a series of lines radiating from the apex down the sides of the pyramid to the base. It is further possible to unfold the surface so that it lies in a flat plane with the lines all radiating from one point which was the apex.

THE RIGHT CONIC FRUSTUM

The right cone is perhaps the most common body, belonging to Class 1, which finds application in sheet metal work.

To obtain the pattern for the frustum of a right cone, first draw the elevation, as at AOL, Fig. 5, and mark off MN at the height of the frustum. With centre O and radius OA, describe the arc $A'A''$ any length. To obtain the length of the perimeter, describe

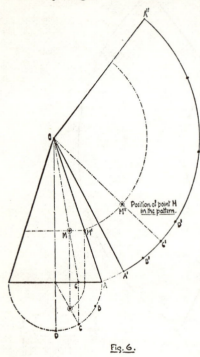

Fig. 6.

one quarter of the base of the cone and divide it into three equal parts, as at $ABCD$. Take one of these divisions, as AB, in the compasses, and mark off three similar distances along the arc $A'B'C'D'$. Four times the distance $A'D'$ will then give the whole perimeter round the arc, as from A' to A''. Join OA''. To complete the pattern for the frustum, draw in the arc $N'P$.

The milk can and the funnel in the top and bottom corners are intended to provide further exercises on this problem.

TRUE LENGTHS

With reference to Fig. 5, it will be seen that the radii of the arcs $N'P$ and $A'A''$ in the pattern are obtained from the slant of the cone, i.e. ON and OA. Distances down the slant of a cone always give true lengths from the apex. Any other lines shown on a cone which do not appear on the outside slant are not true lengths.

For instance, the line OC on the cone shown in Fig. 6 does not represent its true length, but is fore-shortened both in the plan and in the elevation. The true length, however, may easily be obtained by rotating the cone on the centre-point of its base until the point C falls on the slant, as at the position of A. The line OC then coincides with the slant of the cone and its true length is equal to OA.

Furthermore, the true length of any portion of OC may be obtained from its new position on OA.

Take any point M on OC. The distance OM does not represent its true length, but by rotating the cone on its axis the point M may be made to coincide with the point M', when its true length will be seen to be equal to OM'. To obtain its true position on the pattern, locate the position of point C on the perimeter, as at C', and join to the apex O. Swing round an arc from M' until it cuts OC' in M''. M'' is the position of point M in the pattern.

This principle forms the basis of solution to all the problems of the right cone, and should be clearly grasped at the outset.

THE RIGHT CONE CUT OBLIQUELY

Set out the elevation of the cone AOG, Fig. 7, and on its base draw the semicircle ADG. The semicircle thus represents half the perimeter of the base. Divide the semicircle into six equal parts, as at A,B,C,D,E,F,G, and from the points thus obtained draw perpendiculars to the base of the cone, as Bb, Cc, Dd, Ee, Ff. From the points b,c,d,e,f, on the base of the cone, draw lines to the apex O. Draw MN at the position and angle required for the cut-off.

The lines on the cone intersect the plane of the cut-off at $M,1,2,3,4,5,N$; then, to obtain the true distances of these points from the apex, project them horizontally on to the slant of the cone, and from the points thus obtained on the slant, swing arcs round into the pattern. Next, with radius OA, draw in the arc $A'A''$ for the base of the cone and mark off twelve divisions equal to one of those on the perimeter, at as A',B',C',D', and so on. Connect these points to the apex O.

A curve now drawn through the diagonally opposite points of intersection will complete the pattern for the frustum.

The diagrams in the corners should provide additional examples for practice on this problem.

THE RIGHT CONE CUT BY A CURVED OR ANGULAR SURFACE

Set out the elevation of the cone AOG, Fig. 8. From the point d on the centre of the base line describe the arc ST, representing a curved surface, so that it touches the sides of the cone, and would not cut through them if the arc were produced. The lines dS and dT, drawn perpendicular to the respective sides of the cone, will determine the positions of the points S and T. Draw MN at the position and angle required for the cut-off.

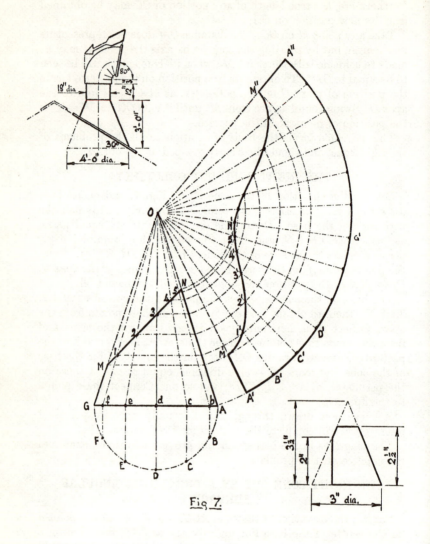

Fig. 7.

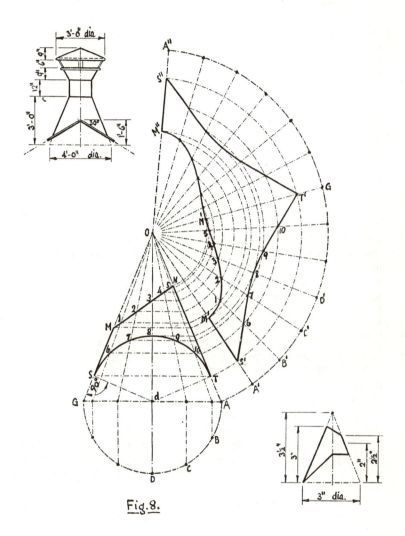

Fig. 8.

On the base of the cone AG describe the semicircle ADG and divide it into six equal parts, as at $A,B,C,D, \ldots G$. From the points thus obtained draw perpendiculars to the base of the cone, and from the points on the base of the cone draw lines to the apex O.

The lines on the cone intersect the curved surface at $S,6,7,8,9,10,T$, and the inclined cut-off at $M,1,2,3,4,5,N$. To obtain the true distances of these points from the apex, project them horizontally, or parallel to the base AG, on to the slant of the cone, and from the points thus obtained on the slant, swing arcs round into the pattern. Next, with radius OA, draw in the arc $A'A''$ for the base of the cone and mark off twelve divisions equal to one of those on the semicircle, as A',B',C',D', equal to AB, or BC, or CD, and so on. Connect the points in the pattern to the apex O.

The pattern may now be drawn in.

From the point M' in the pattern, draw in the curve through the diagonally opposite points $M',1,2,3,4,5,N'$. Repeat the curve to M''. From the point S' in the pattern, draw in the curve through the diagonally opposite points $S',6,7,9,8,10,T'$. Repeat the curve to point S''.

The dimensioned sketches accompanying Fig. 8 provide additional exercises on this problem.

CONICAL JUG TOPS

The developments shown in Fig. 9 are further examples of conic frustums. The jug tops may be developed as right cones cut by a curved, angular, or oblique surface.

The important exercise in connection with this diagram is that of displacing the apex of the cone, as from A to B. Had the lines for the pattern been swung round from the apex A in the ordinary way, the pattern would have encroached badly on the elevation of the jug. In order to swing the pattern clear of the elevation, the line CA and all the points on it should be revolved on the point C to some position CB, such that the complete base curve CDE in the pattern falls clear of the elevation.

This principle may be applied in many cases where the pattern would otherwise encroach on the elevation, and is better than the practice of completely detaching the pattern.

FANCY CANISTER LIDS

The type of canister lid shown in Fig. 10 is a typical example of problems involving the frustum of a pyramid. In the case of a right pyramid, that is a pyramid the axis of which from the centre-point

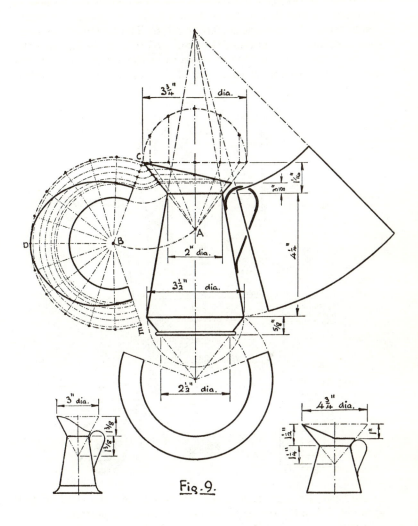

Fig. 9.

of the base to the apex is at right angles to the plane of the base, it should be possible in the plan to draw a circle through all the corner points round the base, as at B,C,D,E,F,G. The centre of the circle then becomes the position of the apex, and lines radiating from the centre to points around the base represent the corners of the pyramid. For the pattern, the diameter of the circle EB should be taken

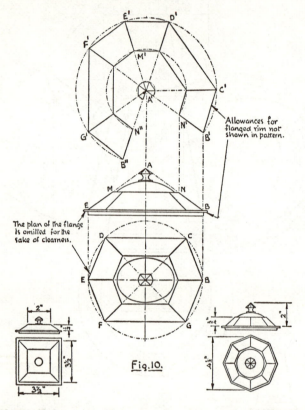

Allowances for flanged rim not shown in pattern.

The plan of the flange is omitted for the sake of clearness.

Fig. 10.

as the base of the pyramid in the elevation, and the slant of the pyramid AB as the radius of the base curve in the pattern.

It will be seen that if the base curve in the pattern were swung out direct from the apex A, the pattern would encroach on the elevation. To avoid this, transfer the line ANB vertically upwards to $A'N'B'$, and swing round the base curve $B'C'D'E'F'G'B''$, and also the inner curve for the top $N'M'N''$. From the plan take the distances BC, CD, DE, EF, FG, GB, and mark these off round the base curve in the pattern as at $B'C'$, $C'D'$, . . . $G'B''$. Lines now drawn from these points to the point of the apex A' represent the corner

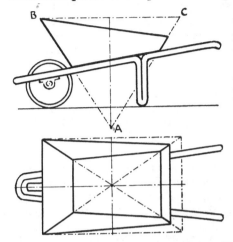

lines in the pattern. Join the corresponding points round the curve for the top, as from N' to M' to N''. If required, the pattern for the top of the knob may be obtained from the same pattern, as shown in the centre.

THE PYRAMID CUT OBLIQUELY

The body of the iron wheelbarrow shown in Fig. 11 is often constructed as the frustum

Fig. 11.

of a pyramid, and may then be developed by the radial line method. The elevation of the complete pyramid is obtained by producing the sides downwards until they meet, as at A, and upwards on the side AC until the

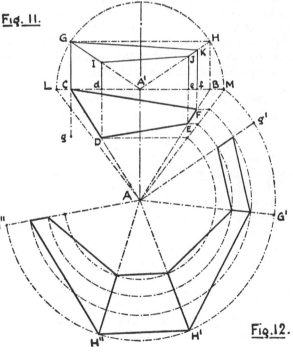

Fig. 12.

base BC is at right angles to, and bisected by, the central axis.

To develop the pattern, set out the elevation as shown at $CDEF$, Fig. 12, and produce the sides to CA and AB. On the base CB draw one half of the plan of the full base, as at $CGHB$. The centre-point A' then represents the apex in the plan. It will be observed that the half-plan and the elevation are drawn the reverse way up to that which is usual, but this should present no difficulty in the solution of the problem if that fact is borne in mind.

From the apex A' draw $A'G$ and $A'H$. From D draw DI; from E draw EJ; from F draw FK, all vertically upwards. Join GK and IJ. Then $CGKf$ is the half-plan of the actual top of the body, and $dIJe$ the half-plan of the bottom of the body. From the apex A' describe the semicircle $LGHM$. Join AL and AM, each of which represents the true length projection of the corner lines $A'G$ and $A'H$. Project the points D,E,F horizontally to the true length line AM, and from A swing them round into the pattern, together with point M.

Take the distance gG, which should be twice the length of CG, and mark off $g'G'$ on the outer, or base, curve in the pattern. Next take GH and mark off $G'H'$ in the pattern. Next mark off $H'H''$ in the pattern equal to $g'G'$. Also mark off $H''G''$ equal to $G'H'$. Join the points thus obtained to the apex A. It now remains to draw in the pattern, which should be evident from the illustration.

THE OVAL

The construction of the oval is a problem which often gives trouble on the score of inaccuracy. The point of doubt arises where the side and end arcs meet, or should meet, and one or two methods, popular with sheet metal workers, need careful observation in this respect. In one case the arcs fail to meet by a distance of one seventy-fifth of the difference between the major and minor axes, which, for ovals of good length compared to the width, is an error of some importance. The method shown in Fig. 13 will be found satisfactory for most purposes, and is easily proved to be true. The arcs will meet at Q if the oval is carefully drawn.

The oval differs from the true ellipse in the fact that the oval is an approximate ellipse drawn with the compasses, and is therefore a construction made up of arcs of circles. On the other hand, no part of a true ellipse is a part of a circle, and cannot, therefore, be drawn with the compasses. The true ellipse will be dealt with later.

CONSTRUCTION 1. Draw AB and CD, the major and minor axes, perpendicular to each other, intersecting at O, the centre. (See Fig. 13.) On one half of the major axis describe an equilateral triangle

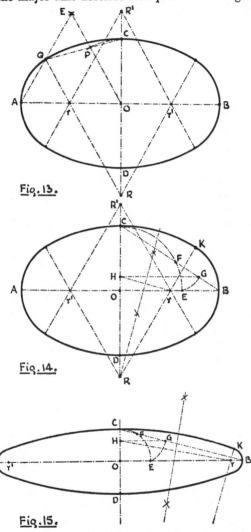

OEA. On OE mark off OP equal to OC. From C, through P, draw CQ. From Q draw QR parallel to OP. Then R is the centre for the arc CQ, and r is the centre for the arc QA. For the corresponding opposite arcs, make OR' equal to OR, and Br' equal to Ar.

Proof. Because OP is equal to OC, the triangle OPC is isosceles, and because QR is drawn parallel to OP, the triangle RQC is also isosceles. Therefore RQ is equal to RC. Again, because QR is drawn parallel to OE, the smaller triangle ArQ is similar to the larger triangle AOE, and is therefore equilateral. Therefore rQ is equal to rA. It will be clear from this that the two arcs AQ and CQ should meet at Q.

Fig. 13.

Fig. 14.

Fig. 15.

When the major axis is considerably longer than the minor axis, nearly 4 to 1, the above method, like many others, fails to give satisfaction, but for general purposes this method will be found very useful, and will be used throughout this work where necessary.

CONSTRUCTION 2. For ovals of all proportions, long or broad, the following method should prove useful. Set out *AB* and *CD*, the major and minor axes, intersecting at *O*, the centre, as in Fig. 14.

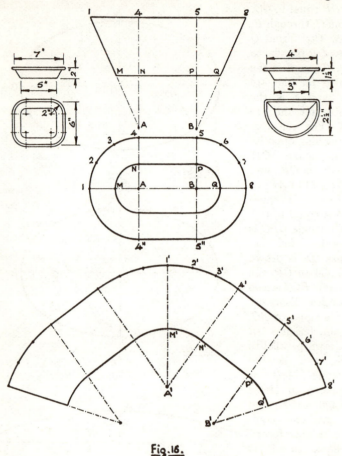

Fig. 16.

Join *CB*. With centre *O* and radius *OC*, describe the arc *CE*, cutting *CB* in *F*. With centre *F* and radius *FE*, describe the arc *EG*. Draw *GH* parallel to *OB*. Take *CH* in the compasses and from *B* mark off *Br* equal to *CH*. Joint *Hr*. Bisect *Hr* and produce the bisecting line to meet *CD* produced in *R*. Then *R* is the centre for the side arc, and *r* is the centre for the end arc. Make *OR'* equal to *OR* for the opposite side arc, and *Or'* equal to *Or* for the opposite end arc.

Proof. In this construction, because the line bisecting *Hr* meets *CD* produced at *R*, then the triangle *RHr* is isosceles, and *RH* is equal to *Rr*. In drawing the arc *CK*, the part *Kr* is cut off equal to *CH*. But *Br* is also cut off equal to *CH*. Therefore *Br* is equal to *Kr*. It will be clear from this that the two arcs *CK* and *BK* should meet at *K*.

This method is not so simple as No. 1, but it is an advantage where the oval is long and narrow, as shown at Fig. 15.

COMPLEX CONICAL PANS

The following problems are typical examples of bodies composed of portions of right cones. The pan shown in the centre at Fig. 16

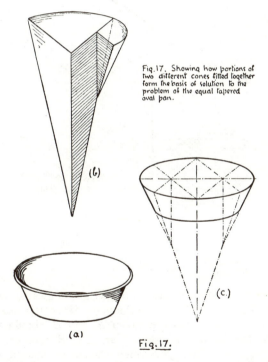

Fig. 17. Showing how portions of two different cones fitted together form the basis of solution to the problem of the equal tapered oval pan.

(b)

(c.)

(a)

Fig. 17.

s made up of two half-conic frustums with two straight flat pieces between them, forming the sides. The two half-cones are shown at *A*,1,4 and *B*,8,5 in the elevation, and at 4,1,4″ and 5,8,5″ in the plan, with the apex *A* on the one side and the apex *B* on the other.

For the pattern, divide half the plan, as shown, by spacing three

equal divisions on the quarter-circle, as at 1,2,3,4, and similar
divisions on the opposite quarter-circle, as at 5,6,7,8. For the
radius of the pattern, take in the compasses the slant length *A*1
from the elevation, and describe an arc from the centre *A'* in the
pattern. Space off 1',2',3',4' equal to the spaces 1,2,3,4 in the

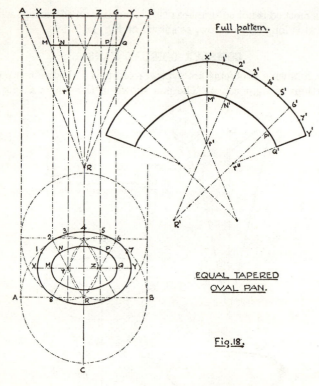

Full pattern.

EQUAL TAPERED
OVAL PAN.

Fig.18.

plan. Join *A'*4'. From 4' draw 4'5' at right angles to the radius
*A'*4' and make 4'5' equal to 4,5 in the plan. Draw *B'*5' parallel and
equal to *A'*4'. The distance between *A'* and *B'* should therefore
be equal to the distance between 4' and 5'. *B'* now becomes the
centre for the arc 5',6',7',8', which should be spaced equal to
5,6,7,8 in the plan. Join *B'*8'. For the inner line, take *AM* in the
compasses from the elevation, and with centre *A'*, describe the arc
M'N' in the pattern. Draw *N'P'* parallel to 4'5'. With radius *B'P'*,
draw the arc *P'Q'*. This completes the half-pattern. For the full
pattern, repeat the process on the other side of the line *A'*1'.

OVAL CONICAL PAN

The equal tapered oval pan shown at (*a*), Fig. 17, is geometrically formed by the combination of portions of two right cones of different dimensions. Portions of two such cones are shown at (*b*), while the figure at (*c*) represents the method of combining portions to form the full pan.

To develop the pattern, set out the plan and elevation as shown at Fig. 18, using the method of drawing the oval as shown at Fig. 13. It will be observed that the point R in this case falls inside the oval. The circle $A4BC$ in the plan (see Fig. 18) represents the base of the major cone, and R is its centre, or apex. The portion of this cone used in the pattern is represented by the section 2–6,R. The base of the minor cone is represented by the circle 8,X,2,Z, and its centre, or apex, occurs at r. In the elevation the major cone is shown at R,A,B, and the minor cone at r,X,Z. In the plan, divide the arcs X–2 and 6–Y into equal parts, as at X,1; 1,2; and 6,7; 7,Y. Divide also the arc 2–6 into equal parts, as at 2,3; 3,4; 4,5; 5,6.

Take in the compasses the slant length rX from the elevation of the minor cone, and describe an arc X'–2′ in the pattern. Mark off X',1′,2′ in the pattern equal to X,1,2 in the plan. Join 2′,r', and produce to R'. Cut off 2′,R', equal to the slant length R,A, in the elevation. Then R' becomes the centre for the major cone pattern. With R' as centre, describe the arc 2′–6′ and space off 2′,3′,4′,5′,6′ equal to the corresponding divisions in the plan. Join 6′,R'. Cut off 6′,r'', equal to 2′,r', on the opposite side. Then r'' becomes the centre for the remaining portion of the minor cone. With centre r'', describe the arc 6′–Y', and space off 6′,7,Y', equal to 6,7 and 7,Y, in the plan. Join Y',r''. For the inner line, take r,M in the compasses from the elevation, and with centre r', describe the arc $M'N'$ in the pattern. With centre R', describe the arc $N'P'$, and with centre r'', describe the arc $P'Q'$.

This completes the half-pattern. For the full pattern repeat the process on the other side of the centreline $X'r'$.

CHAPTER 3

THE PARALLEL LINE METHOD

THE Parallel Line method of development, as described in the chapter on Classification of Geometrical Forms, is based on a system of lines drawn parallel to one another on the surface of a solid. For this condition to hold good, the solid must belong to the class of prisms which preserves a constant cross-section, equal in shape and size, throughout the length. Pipe work and duct systems usually furnish an abundance of geometrical problems involving the parallel line method of development, as also do many forms of chutes, spouts, and ornamental mouldings. In some respects the principles of the Parallel Line method are the easiest to grasp in the early stages, but many examples in the more advanced grade present intricate problems of intersections before the patterns can be developed. Perhaps the simplest example illustrating this method is afforded by a square or rectangular pipe, in which case the corners represent the parallel lines on the surface.

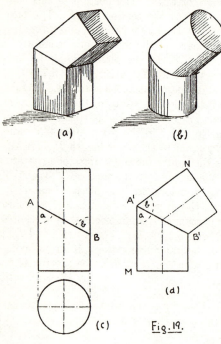

Fig. 19.

ELBOWS

An elbow in pipe work is a single angular diversion from a straight line. Fig. 19 (a) shows a rectangular pipe elbow and (b) a corresponding cylindrical pipe elbow. The position of the joint line, shown at $A'B'$, Fig. 19 (d), bisects the angle of the elbow. This will be seen

24

from the illustration at (c). The figure represents a short piece of straight cylindrical pipe, cut through diagonally at AB. This makes the angle at (a) equal to the angle at (b). If, now, the top piece is reversed so that the point A falls on the point B, then an elbow, as shown at (d), is obtained, in which the joint line $A'B'$ bisects the angle $MA'N$.

A plain elbow cannot be formed between pipes of unequal diameters, because the ellipses formed by the angular cut-off on each pipe will not fit. Fig. 20 (a) shows two such pipes cut off at equal angles L, L. The pipes, clearly, do not match up to form an elbow. The illustration at (b) is not quite so obvious, and represents a pitfall which many fall into in the early stages. In this case the angles at M and N are not equal. The joint line AB, therefore, does not bisect the angle of the elbow. Furthermore, although the line AB represents the major axis of both the ellipses formed by the two pipes, the ellipses are not equal, since the

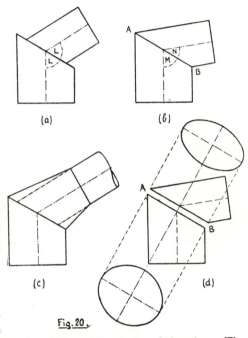

Fig. 20.

minor axes correspond to the different diameters of the pipes. The figure at (d) shows the projected ellipses as would occur at AB.

An elbow may be formed with two pipes of unequal diameters by inserting a conical connecting piece between them, as shown at Fig. 20 (c). This problem will be dealt with later.

THE SQUARE PIPE ELBOW

To develop the pattern for the square pipe elbow shown at Fig. 21, imagine the pipe to be unfolded or unrolled at right angles to its central axis. Draw the base line $S'S''$ and mark off $S'1'2'3'4'S''$ equal to the corresponding distances round the perimeter, or girth line, in the plan. From the points thus marked on the base line erect perpendiculars parallel to each other. From the points A and B in the elevation draw lines parallel to the base line in the pattern,

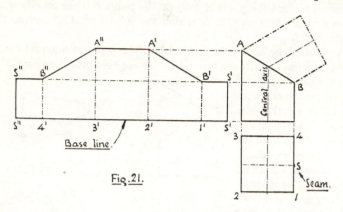

Fig. 21.

intersecting the perpendiculars in $S'B'A'A''B''S''$. The outline of the pattern should now be evident from the illustration given. The same contour $S'B'A'A''B''S''$ should serve for the other side of the elbow.

THE CYLINDRICAL PIPE ELBOW

The principle of setting out the pattern for a cylindrical pipe elbow is much the same as for the square pipe elbow. The chief difference

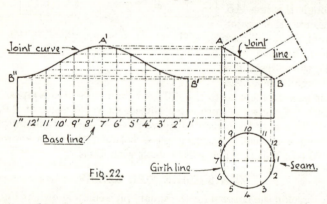

Fig. 22.

is that, as there are no natural corners, the circular girth line, shown in the plan, is divided up into any convenient number of parts—twelve here shown—and numbered accordingly, as in Fig. 22.

Set off the base line in the pattern, and mark off the divisions 1′2′3′ . . . 12′1″, equal to those round the girth line. From these points on the base line erect perpendiculars parallel to each other. Next, from the points around the girth line in the plan, draw lines vertically upwards to cut the joint line AB in the elevation, and from these points of intersection on AB, draw lines horizontally into the pattern to meet the perpendiculars from the base line. A curve drawn through these points in the pattern should give the contour of the joint line, and the full pattern may be completed.

THE DOUBLE RETURN ELBOW

The pattern for the middle section of the double elbow shown at Fig. 23 is really equivalent to two single elbow patterns back to back.

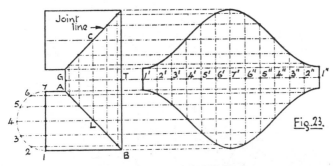

Fig. 23.

First, draw the semicircle on the end of one section, as shown, which represents half the circumference, or girth of the pipe. Divide the semicircle into six equal parts and number the points from 1 to 7. From these points project lines perpendicularly back to the end of the pipe, and on to the joint line AB. From the points where these lines intersect the joint line AB, draw lines parallel to the centreline CL of the middle section, across to the opposite joint line.

Now take a girth line GT at right angles to the centreline CL of the middle section, and extend it to any length into the pattern. On this extended girth line mark off the divisions 1′2′3′4′5′6′7′ . . . 1″ equal to twice those round the semicircle on the end of the pipe. The full distance from 1′ to 1″ should therefore be equal to the full circumference of the pipe. Through these points draw lines parallel to each other at right angles to the girth line.

Next, from the points on both joint lines of the middle section, draw lines parallel to the extended girth line to meet those at right angles to it. Curves drawn through those points of meeting or crossing should give the joint curves of the pattern. The pattern may now be completed by drawing in the short straight lines at the ends.

THE DOUBLE OFF-SET ELBOW

After the previous problem it should not be difficult to develop the pattern for the middle section of the off-set elbow, as shown at Fig. 24, without a full description.

One particular point should be noted, however, and that is that

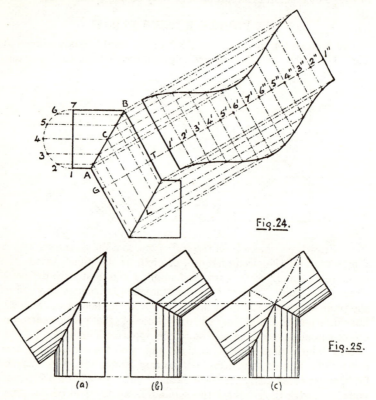

Fig. 24.

Fig. 25.

(a)   (b)   (c)

the girth line GT and its extension into the pattern is again at right angles to the centreline CL of the middle section.

TEE JOINTS AND BRANCHES

A tee joint may be regarded as a combination of two elbows. Fig. 25 (a) shows an acute elbow, and (b) an obtuse elbow, in which

the combined angles form 180 degrees. The figure at (c) shows the
two elbows combined, forming an oblique tee joint, This illustration,
in conjunction with Fig. 19 (c), may serve to explain why the joint
lines in tees between pipes of equal diameters are always represented
by straight lines in the elevation. All cases of tees between pipes of
unequal diameters present curved joint lines.

THE RIGHT-ANGLED TEE

To develop the pattern for the branch pipe forming a right-angled
tee joint, as shown at Fig. 26, first draw the semicircle on the end of

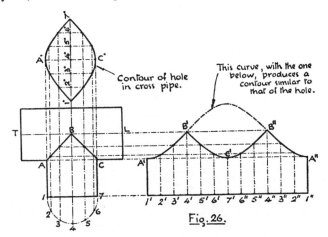

Fig. 26.

the branch pipe, which represents half the circumference or girth
of the pipe. Divide the semicircle into six equal parts and number
the points 1 to 7. From these points project lines perpendicularly
up to the end of the pipe, and on to the joint line ABC. Next, set
off the base line in the pattern, and mark off the divisions $1'2'3'4'5'6'$
$7'$ . . . $1''$, equal to those round the semicircle. From these points
on the base line, erect perpendiculars parallel to each other. Now,
from the points on the joint line ABC, draw lines horizontally into
the pattern to meet the perpendiculars from the base line. Curves
drawn through these points in the pattern, as shown in the diagram
through $A'B'C'B''A''$, should give the true form of the joint line.
The full outline of the pattern may now be completed.

The shape of the hole in the cross pipe may be developed by
projecting the points on the joint line ABC, in the elevation, upwards
at right angles to the centreline TL. On the middle line, mark off

distances equal to those on the semicircle, as at 1',2',3',4',5',6',7'. Through these points draw lines at right angles to those projected upwards, and draw in the curves through the points of meeting, as shown in the diagram.

Incidentally, if the curve $B'C'B''$ in the pattern be repeated on the opposite side of the line $B'B''$, a similar contour of the hole will be obtained.

THE OBLIQUE TEE

The examples shown at Fig. 27 represent typical problems of the stove pipe and smoke cowl. The method of developing the pattern for the oblique joint at the bottom is very much the same as for the

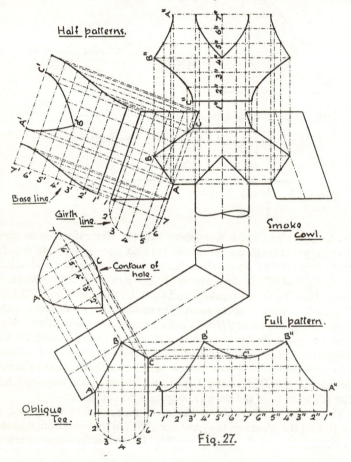

Fig. 27.

right-angled tee given in the previous exercise, and should be readily followed by reference to the diagram.

The smoke cowl above represents a combination of tee joints, and, whilst the principles of development remain the same, the example provides additional exercises which are well worth following up. In the illustration only half-patterns are shown for the middle section and end pieces. One point of importance should be observed, and that is that the semicircle representing the girth line is drawn at right angles to the central axis of the left end piece. Furthermore, the base line for the pattern is also projected in line with the diameter of the semicircle, at right angles to the central axis of the end piece.

THE RIGHT AND OBLIQUE CYLINDERS

The ordinary process of rounding up or rolling a pipe makes it circular in cross-section, but there are times when a pipe must be oval and not circular, as, for example, when a connecting piece is required between two circular holes which are in parallel planes, but not in perpendicular line with each other. Sometimes these cases prove troublesome, not because of any special degree of intricacy, but generally through failure to appreciate the simple geometrical difference between the right cylinder and the oblique cylinder. The one can so easily be mistaken for the other unless careful observation is exercised in noting the particulars given. For instance, the branches forming the "Y" piece shown in Fig. 31 might be taken for ordinary round or cylindrical pipes, but actually the pipes are oval in cross-section.

A very simple method of demonstrating the geometrical difference between these two cylinders is to take a number of pennies and place them one above another in an upright pile. If the pile stands vertically on a horizontal surface,.as shown in Fig. 28 (a), then a right cylinder is represented. Suppose now that some small block be placed under the base at one side of the pile, causing the whole to lean over as shown in (b): this does not alter the geometrical properties of the pile itself. It is merely inclined at an angle to the vertical. Each penny is still at right angles to the central axis, and the pile still represents a right cylinder.

Let the small block now be removed so that the pile may return to its original position. If, now, each penny be pushed slightly forward, sliding on the one below, so that the pile as a whole, but not each penny, leans forward forming an angle with the vertical, as shown in Fig. 28 (c), then an oblique cylinder is represented. In this case it will be observed that the pennies do not lie at right angles to the central axis of the pile, as they do in the right cylinder. The figure at (d) shows the oblique pile tilted into the vertical position, but the geometrical conditions remain the same.

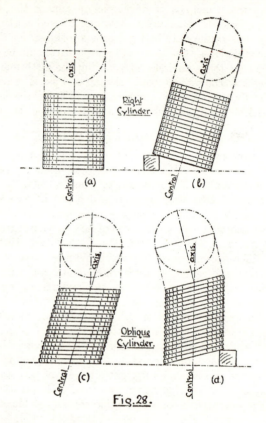

Fig. 28.

DEFINITIONS. A Right Cylinder has its central axis at right angles, or perpendicular, to the plane of its circular base.

An Oblique Cylinder has its central axis inclined at an acute angle to the plane of its circular base.

These conditions, it may be noted, are similar to those governing the Right and Oblique cones, with the difference that the cones taper from the base to an apex, whereas the cylinders preserve equal cross-sections throughout their lengths. It is important to grasp fully the geometrical difference between these two cylinders as on this depends the success of many solutions involving the cylinder.

THE RIGHT CYLINDER CUT OBLIQUELY

The two problems shown in Figs. 29 and 30 illustrate the difference in the patterns for a right cylinder and an oblique cylinder, as represented by similar elevations. The illustration at Fig. 29 is a right

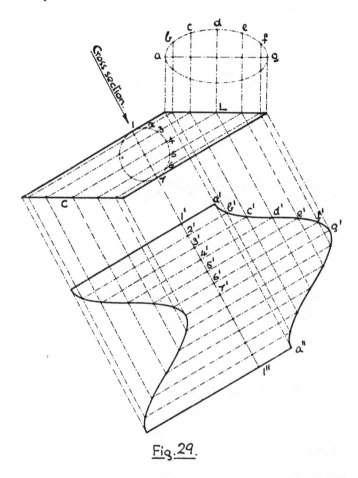

Fig. 29.

cylinder cut obliquely at each end. It is therefore circular in cross-section, as shown by the circle on the centreline, but the shape at both ends is elliptical, as shown by the projection at the top. The development of the pattern is obtained in a similar manner to that of the middle section of the double off-set elbow, Fig. 24. The circumference of the pipe is therefore spaced along the line 1'1″, at right angles to the centreline *CL* of the cylinder.

A point of special interest may here be noted, particularly in view of the problem to follow. The spacings *a,b,c,d,e,f,g* around the ellipse

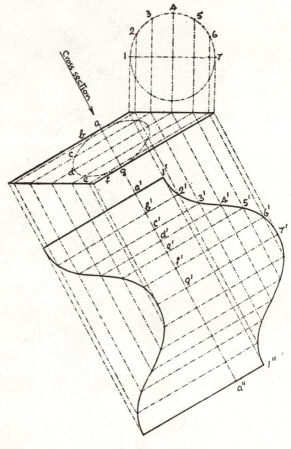

Fig. 30.

are similar and equal to those at *a′,b′,c′,d′,e′,f′,g′* around the curve in the pattern.

THE OBLIQUE CYLINDER

In the diagram Fig. 30 the shape of the cross-section is elliptical, as shown by the ellipse on the centreline. But in this case the ends are circular, as shown by the circle projected above at the top. This condition establishes the fact that the cylinder is truly oblique, and not, as in the previous example, a right cylinder cut off obliquely. In developing the pattern, one of two alternatives may be adopted to obtain the girth of the cylinder. Either the unequal spacings a,b,c,d,e,f,g around the ellipse may be marked off along the line $a'a''$, or the equal divisions 1,2,3,4,5,6,7 around the circle may be marked off, stepping from one line to the next, to form the curve $1',2',3',4',5',6',7'$ in the pattern. Generally the latter is the more convenient, as the compasses may be set to one division only, and the whole distance spaced off right away.

THE OBLIQUE CYLINDRICAL " Y " PIECE

The two "Y" pieces shown in Figs. 31 and 32 are alternative constructions involving the principles of the oblique and right cylinders.

To develop the pattern for the oblique construction shown in Fig. 31, describe a semicircle on the end of one branch of the "Y"

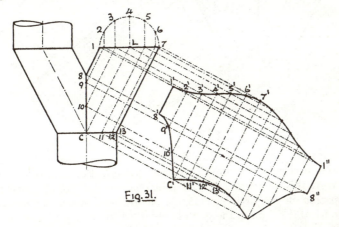

Fig. 31.

piece, and divide into six equal parts, as shown at 1,2,3,4,5,6,7 From these points drop perpendiculars to the end of the branch and from these draw lines parallel to the central axis CL to cut the joint lines at the other end, as at $8,9,10,C,11,12,13$. From the points on the joint lines at both ends, project lines at right angles to the central axis into the pattern. To space out the girth of the pipe, take in the compasses one of the equal divisions from the semicircle, as from 1 to 2, or 2 to 3, and, beginning at any convenient spot on the line projected from point 1, space off $1',2',3',4',5',6',7' \ldots 1''$, by stepping the distances from one line to the next. A curve drawn through these points will give the outline of the top end. Next, from the points on this curve, draw lines at right angles to those on which the points are spaced, to cut the lines projected from the other end of the pipe, as at 8′,9′,10′,C′,11′,12′,13′ . . . 8″. Curves drawn through these points, as shown in the diagram, will give the outline of the bottom joint line. To complete the pattern draw in the two end lines 1′,8′ and 1″,8″.

THE RIGHT CYLINDRICAL " Y " PIECE

In the right cylindrical construction of the "Y" piece, as shown in Fig. 32, the joint lines bisect the angles of the central axis, as in

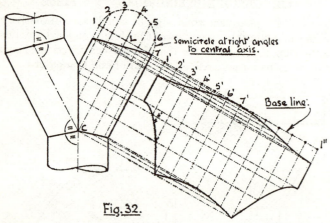

Fig. 32.

the case of ordinary elbows. This ensures the cross-sections being equal and circular. In developing the pattern, the chief difference between this problem and the previous one lies in the method of spacing off the girth line. It will be observed that in the oblique cylindrical "Y" piece the girth is obtained indirectly by the alternative method of spacing the distance round the curve. But in the example shown in Fig. 32, the semicircle from 1 to 7 represents the girth or circumference of the pipe, and the spacing in the pattern is made direct along the base line.

For the remainder, the procedure is similar to that of the previous example.

THE SUBCONTRARY SECTION

An important property of the prism, and therefore of the cylinder, is that the shape of any section across its central axis has its counterpart at an equal angle on the opposite side of a perpendicular through the central axis. This will be seen by reference to the diagrams (a), (b), (c), Fig. 33, which represent oblique cylinders leaning at different angles. The bases, therefore, are circular, and any cross-section parallel to the bases AB will also be circular.

But another circular cross-section occurs at any plane which crosses the central axis at an equal, but opposite, angle to AB. Thus at (a), Fig. 33, if the plane of the base AB be revolved about any point O in AB produced, until the point A falls on the point D, then the angle N will be equal to the angle M, and CD equal to AB.

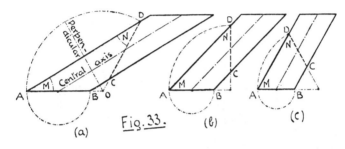

Fig. 33.

(a)   (b)   (c)

But CD represents the position of such a plane as will produce a similar cross-section to that of AB. The shape is therefore circular. This alternative position may be termed the Subcontrary section, since it is similar, but opposite, to the base section.

This principle is a very important one, and is a deciding factor in the method of solution applied to many parallel line problems. The middle diagram (b) is perhaps the most important, as its central axis leans at 45 degrees, which brings the base and its subcontrary section into an angle of 90 degrees.

AN OBLIQUE CONNECTING PIPE

Fig. 34 represents a connecting pipe between two circular holes in planes at 90 degrees to each other. The holes are so placed as to make the connecting pipe lean at 45 degrees, thus forming equal angles with the holes at each end.

By these conditions the connecting pipe is an oblique cylinder. To develop the pattern, project a centreline at right angles to the central axis, CL, through its centre-point O. Draw the semicircle on the base line AB, to represent half the plan of the base end. Divide the semicircle into six equal parts, as at $A,1,2,3,4,5,B$; and from these points project lines perpendicularly back to the base line. From the points obtained on the base line AB, draw lines into the pattern at right angles to the central axis CL. From any point B' on the line projected from B, mark off $B',5',4',3',2',1',A',\ldots$ B'', equal to the corresponding distances round the semicircle. The spacings should be stepped over from one line to the next, until the outside point A' is reached, and then repeated inwards to point B''.

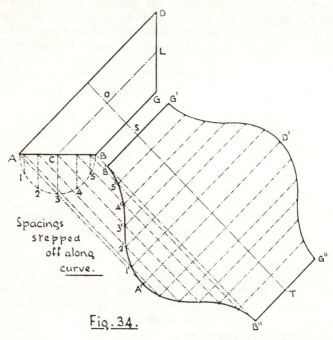

Fig. 34.

A fine drawn through these points will give the form of the base curve in the pattern. From these points on the base curve, draw lines parallel to the central axis CL, or at right angles to the centreline ST of the pattern, and on each mark off an equal distance on the other side of the centreline ST. A line drawn through this new set of points will give the opposite end curve. The two end lines $B'G'$ and $B''G''$ complete the pattern.

THE LOBSTER-BACK BEND

There is probably no pattern in sheet metal work so much used as that of the lobster-back bend. It is generally regarded as one of the simplest to set out, which, in some respects is true, but the

method of development is by no means free from the confusion on
the difference between the right and oblique cylinders. There are
two methods in general use of setting out the elevation of a six-
segment bend. In one method the segments are portions of oblique

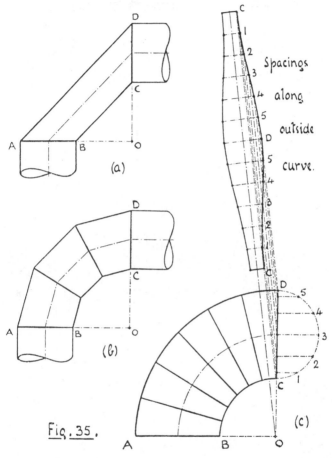

Fig. 35.

cylinders; in the other they are portions of right cylinders; yet
the corresponding methods of development are often applied indis-
criminately, sometimes, naturally, with disappointing results.

Referring to Fig. 35 (c), it will be seen that the bend is composed
of six full segments between the arms of the right angle *AOD*. The
segments are, therefore, parts of oblique cylinders. This conclusion

may be arrived at by an inspection of the diagrams at (*a*) and (*b*), Fig. 35. At (*a*) it will be seen that there is only one connecting piece between the circular pipes *AB* and *CD* terminating on the arms of the right angle *AOD*. This connecting piece is unquestionably an oblique cylinder, similar to that of Fig. 34. The diagram at (*b*) shows

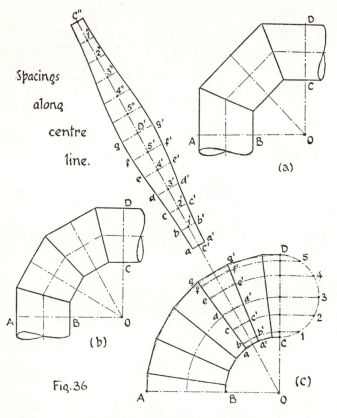

Spacings along centre line.

(a)

(b)

Fig. 36

(c)

the two pipes *AB* and *CD* connected by a series of three such pieces. In this case each segment is similar to the cylinder *ABCD*, shown at (*c*), Fig. 33, which again is oblique. Similar reasoning will show that each piece of the six-segment bend at (*c*), Fig. 35, is also an oblique cylinder. The method of developing the pattern for one segment is, therefore, exactly the same as that shown at Fig. 34, and the development at (*c*), Fig. 35, should be easily followed from this without further explanation.

Referring now to Fig. 36 (c), it will be seen that the bend is composed of five full segments, and half a segment at each end, making six in all. The segments of this bend are portions of right cylinders, as may be gathered from an inspection of the diagrams at (a) and (b), Fig. 36. At (a) only one connecting piece is shown between the two pipes AB and CD, but these two pipes are extended to form right cylindrical elbows at each joint, similar to the example shown at Fig. 22. Between the arms of the right angle AOD there are, therefore, in addition to the middle connecting piece, two half-segments, one at each end. The diagram at (b) shows the bend composed of two full segments and a half at each end. The bend shown at (c) contains five full segments and two halves.

In this type of construction the half-segment at each end usually forms an extension of the main pipe itself, thereby saving a joint. Moreover, the girth or circumference of each segment is circular and equal to that of the main pipe. The method of development in this case should be similar to that shown at Fig. 23, wherein the girth of the pipe is spaced along the centreline of the pattern, as, in this case, at $C',1',2',3',4',5',D', \ldots C''$. The width of the pattern at each point should be equal to the corresponding length of the straight line across the segment, as $aa',bb',cc',dd',ee',ff',gg'$.

This method of development is often applied to the type of lobster-back bend shown at Fig. 35, but, whilst the actual difference in the patterns for six-segment bends is hardly noticeable, it will be found that, for bends of only three segments, the difference in the patterns will be a source of considerable trouble at the end joints, because the length of the curve on the pattern will be much longer than the circumference of the pipe to which it must be joined.

THE ELBOW GUSSET PIECE

There are many forms of elbow gusset pieces which serve the useful purpose of strengthening the joint between two pipes. In these problems, too, it is important to distinguish between constructions involving the principles of the right cylinder and those based on the oblique cylinder. At (a), Fig. 37, the gusset piece is a portion of a right cylinder, with two flat triangular pieces at each side. It is a portion of a right cylinder because the joints at CA and LB bisect the angles between the centrelines; thus angle M is equal to angle N.

To develop the pattern, from the point O project a centreline at right angles to the central axis CL. From the point R, where the centreline crosses the central axis, describe the quadrant 1–4, and divide it into three equal parts, as at 1,2,3,4. Through these points draw lines parallel to the central axis *CL* to cut the joint lines *CA* and *LB* at each end. From the points on the joint lines at both ends project lines into the pattern at right angles to the central axis *CL*.

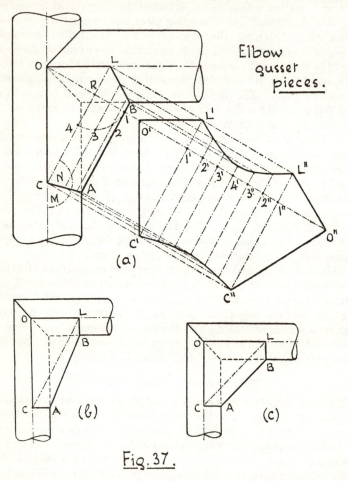

Elbow gusset pieces.

Fig. 37.

On the centreline projected through *OR* into the pattern, mark off the distances 1′,2′,3′,4′,3″,2″,1″, equal to the divisions 1,2,3,4, round the quadrant. Through the points on the centreline *O'O″*, draw lines parallel to the central axis *CL*, or at right angles to the centre-line, to cut the lines projected from the joints *CA* and *LB*. Draw curves through the points of intersection, as from *L'* to *L″*, and from

C' to C''. To complete the pattern, by adding the triangles at each end take the distance RO in the compasses, and from the points $1'$ and $1''$ in the pattern, mark off the distances $1'O'$ and $1''O''$. Join $L'O'$ and $C'O'$, also $L''O''$ and $C''O''$.

The diagram at (c), Fig. 37, represents a gusset piece in which the portion $CLBA$ is a part of an oblique cylinder. It is similar to the portion $CLGB$ of the oblique cylinder shown at Fig. 34, and should be developed in a similar manner. The two flat triangles CLO should then be added, as in the previous example. The diagram at (b), Fig. 37, represents a gusset piece which cannot be developed by the parallel line method, and any error in that connection should be carefully guarded against. The central axis CL is not parallel to the line AB, and the problem does not conform to the usual method of solution. It may, however, be dealt with by triangulation, but it would be much better to alter the details of construction to bring it into line with the problem shown at (a).

AN OPEN CHUTE

The illustration of the chute in Fig. 38 is another common example of the application of the oblique cylinder in practice. It represents a delivery chute, or spout, from the outlet of a mixer. The bottom part of it, below the central axis CL, is half an oblique cylinder, because the semicircular end at $L5$ is inclined at an acute angle to the central axis. The front elevation on the right is a view looking into the spout in the direction of the arrow.

The pattern is most conveniently developed from the side elevation on the left, by placing half of the contour of the top edge from the front elevation, as from 1 to 5, against the corresponding edge $TL5$ in the side elevation. Thus the quadrant $L,2,5$ is equal to the quadrant $O,2,5$, and the vertical lines 1,2 in both views are equal. Divide the quadrant $L,2,5$ into three equal parts and number the points 2,3,4,5. From these points project lines perpendicularly back to the edge $L5$, and from the points on $L5$ draw lines parallel to the central axis CL, to cut the other edge at the bottom end of the chute. From the points on both edges project lines into the pattern at right angles to the central axis. Next, take the distance 1,2 in the compasses, and from any point $1'$ on the line projected from T, mark off $1'2'$ by stepping over from one line to the next, as shown in the diagram. Follow this up by spacing off $2',3'$; $3',4'$; $4',5'$; equal to the corresponding spaces round the quadrant. Repeat these spacings back to the point $1''$. A line drawn through these points will then give the form of the top curve in the pattern. From the

points $2', 3', 4', 5', \ldots 2''$ on the curve, draw lines parallel to the
central axis CL to cut the lines projected from the other end of the
chute. A similar curve drawn through the points of intersection,

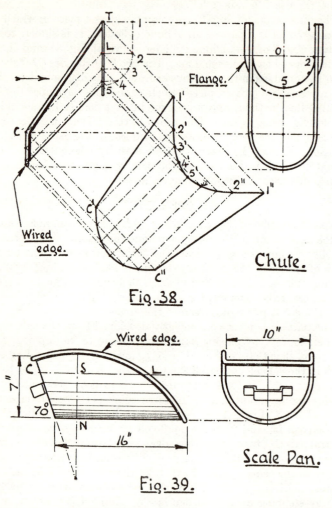

Chute.

Fig. 38.

Scale Pan.

Fig. 39.

as from C' to C'', will give the form of the bottom curve in the pattern
Join $1', C'$, and $1'', C''$. It will be observed that the two triangles,
$1', 2', C'$ and $1'', 2'', C''$, are extra to that portion of the pattern which
forms the half oblique cylinder. When the chute is rolled into shape,

these two triangles should remain flat. No allowances for the wired edge are shown in the pattern. These should, of course, be added to suit the gauge of wire used.

The scale pan shown in Fig. 39 is an example involving the principle of the right cylinder, because the semicircular cross-section occurs at *SN*, at right angles to the central axis *CL*. The development of the pattern, however, is left as an additional exercise.

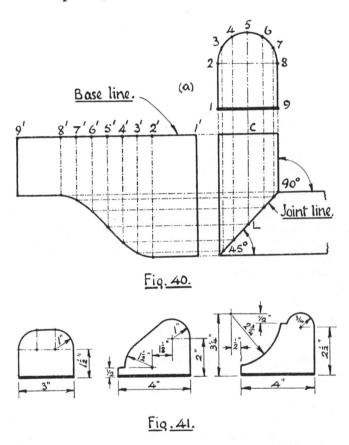

Fig. 40.

Fig. 41.

MOULDING: CURBS AND GUTTERS

The geometry of moulding sections should offer no difficulty if it be remembered that these are ordinary prisms, and the methods of development are precisely the same as those of the cylinder, except that the shape of the cross-section is varied instead of circular. There is no difference between a curb and a gutter in so far as their geometry is concerned. In a practical sense, the difference lies in the use to which they are put; the gutter for roofing work; the curb for the hearth. The illustration at (a), Fig. 43, represents the cross-section of a curb. If this diagram be reversed, turned upside-down, a typical form of gutter will be seen. The development of the pattern is therefore exactly the same for each. Curbs offer scope for a great variety of treatment, not only in design, but also in the practical construction and the choice of material. However, those sections illustrated in Figs. 40 and 41 are suitable for making of 18 or 20 gauge sheet-brass or, alternatively, of mild steel, welded joints, to be afterwards suitably enamelled in a variety of colours.

To develop the pattern for the mitred joint, first draw the shape of the cross-section, as shown at (a), Fig. 40, and immediately below it the plan of the elbow. Divide the contour of the cross-section into any convenient number of parts, as shown at 1,2,3,4,5,6,7,8,9, and from the points of division, drop vertical lines into the plan, to cut the joint line below.

From the points on the joint line, draw lines into the pattern at right angles to the central axis CL. Next, set off the base line at right angles to the central axis, and mark off the divisions 1',2',3',4',5', 6',7',8',9', equal to the corresponding divisions round the cross-section above. From these points on the base line, drop perpendiculars to cut the lines projected from the joint line in the plan. A line now drawn through the points of intersection will give the contour of the joint line in the pattern. The same contour reversed serves for the pattern of the other side of the elbow.

The angle of the joint line, more particularly in connection with gutters and spouting, makes no difference to the method of procedure. The examples given show the joint line of 45 degrees, which forms an elbow of 90 degrees. For angles which are required to be greater or less than 90 degrees, all that is necessary is to mark the joint line at an angle which is half of that of the desired elbow. Thus, if the elbow is required to be 120 degrees, the angle of the joint line should be at 60 degrees, or the angle of the elbow bisected. Then proceed as in the example given in Fig. 40. The sections shown in Fig. 41 serve as additional examples for practice.

The contour of the cross-section also makes no difference to the method of procedure, as may be seen by an inspection of the development shown in Fig. 43. The curb illustrated in Fig. 42, with the development of its mitred joint shown below, is an example of one made of 18 gauge brass, with panels of repousséd design along the front and sides.

THE WHEEL-ARCH PANEL

The illustration in Fig. 44 is a typical example of the wheel-arch panel of a motor car. The development of the pattern for the blade

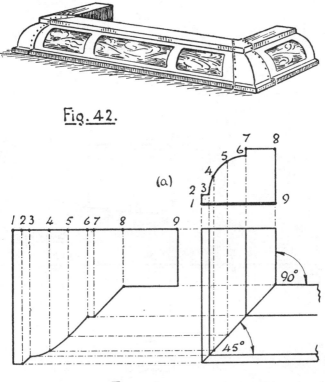

Fig. 42.

Fig. 43.

is a straightforward problem of the parallel line method, and though the details of design may vary, the general principle of the development should apply in all such cases.

In the front elevation, the edge of the blade is divided equally into fifteen parts, and numbered 1 to 16. From these points, lines are dropped vertically into the plan to form parallel lines across the width of the blade. To develop the pattern, set off the base line, shown below, and mark off the distances 1',2',3',4' . . . 16', equal to those spaced round the edge of the blade in the elevation. From the points thus marked off, project lines at right angles to the base line into the pattern. Next, on the projected lines, mark off the corresponding lengths from the parallel lines across the blade in the

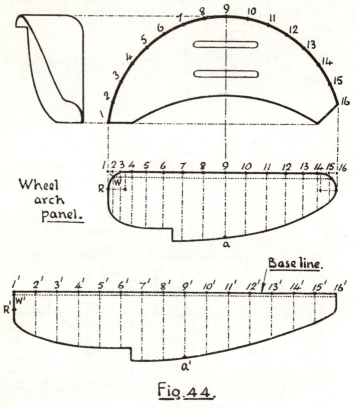

Wheel arch panel.

Base line.

Fig. 44.

plan. Thus $9', a'$ in the pattern should be equal to $9, a$ in the plan. A curve through these points will give the shape of the pattern.

An important consideration is the position of the welded joint between the blade and the shield. This is usually made along the centre of the curved corner, as shown by the dotted line in the plan. To obtain the position of this line in the pattern, mark off the distance $1'R'$ in the pattern equal to $1R$ from the plan. Then, from R' in the pattern, mark back $R'W'$ equal to the curved portion RW from the plan. The dotted line in the pattern thus represents the position of the weld, and the strip should be cut off to that line.

CHAPTER 4

THE METHOD OF TRIANGULATION

TRIANGULATION is by far the most important method of pattern development, since the greater part of the geometry of sheet metal work is concerned with bodies of complex design. Triangulation is often called the "Short Radius" method, but the choice of terms matters little beyond the fact that the former is the more appropriate. In a broad sense, triangulation depends on a method of dividing the surface of the object into triangles, finding the true size of each triangle separately, and placing them side by side in the proper order to obtain the full pattern. It is, therefore, a process of addition, or building up. To obtain the true size of each triangle, the true length of each side must be found and placed in its correct relation to the other sides.

THE GOLDEN RULE OF TRIANGULATION

The method of finding the true length of a line is very simple, yet sometimes very elusive in a complex problem. This method, sometimes called the "Golden Rule" of triangulation, is to place the plan length of a line at right angles to its vertical height, when the diagonal will represent its true length. The principle of this rule will be seen by reference to Fig. 45, which shows, pictorially, a ladder leaning against a wall.

The PLAN LENGTH will be the horizontal distance from the foot of the wall to the foot of the ladder; the VERTICAL HEIGHT will be the height up the wall to the point where the top of the ladder leans against it; and the TRUE LENGTH will be the actual length of the ladder itself. It will be observed that when the position of the ladder is altered the plan length and the vertical height vary accordingly, but when placed at right angles to each other always produce a true length diagonal.

In triangulation two views of an object are essential: an elevation and a plan. Since the elevation is a view in a vertical plane, corresponding to the wall in Fig. 45, it follows that the vertical height of any line may be found from the elevation. Since the plan is a view in a horizontal plane, corresponding to the ground level in Fig. 45, it follows that the plan length of any line may be found from the plan. All that is necessary, therefore, is to take the length of any particular line from the plan, and place it at right angles to the vertical height of the same line from the elevation, when the diagonal will give the true length. (See Fig. 46.)

This sounds simple enough, but if skilful mastery of triangulation

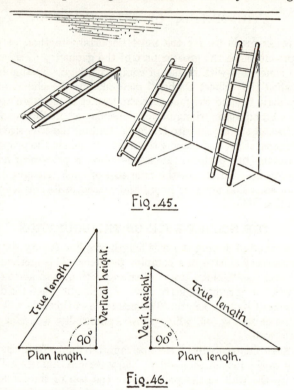

Fig. 45.

Fig. 46.

is to be gained, careful observation must be exercised in dividing up the surface of the object into triangles and in following the correct routine of development.

DIVIDING THE SURFACE INTO TRIANGLES

It will be remembered that, in the chapter on Classification of Geometrical Forms, a piece of cord was used as a medium for outlining the surface of the solid, and that, in the case of those solids belonging to the class of transformers, the cord, whilst kept taut, was made to trace out different forms at each end, such as a square at one end to a circle at the other, or a rectangle at one end to an ellipse at the other. It is necessary now to recall an important factor in this illustration. The use of the piece of core, even though in imagination, is a valuable aid to the correct division of the surface of the object into triangles. The cord must lie in a straight line along the surface of the object from end to end, or as eeaily as the form will permit.

The importance of this will be seen by reference to Fig. 47. At (*a*) the illustration represents, perspectively, a square-to-circle transformer, commonly called a tallboy. It is shown in orthographic plan and elevation below at (*b*). This type of geometrical body is very common in ductwork, and serves to transform a square or rectangular section to that of a circular or cylindrical form.

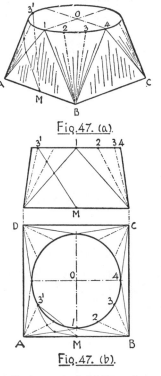

Fig. 47. (a).

Fig. 47. (b).

The surface of this transformer is composed of four separate quarters of a cone, with flat triangles between them. The quadrant numbered 0,1,2,3,4 represents the base of one quarter-cone, and the corner point *B* represents the apex. The triangles 1*AB* and 4*BC* are flat surfaces between this quarter-cone and the two quarter-cones on either side. A piece of cord, stretched taut, with one end fixed at *B*, could be made to lie in a straight line along the surface of the quarter-cone at any position between 1 and 4. Therefore the two intermediate lines *B*2 and *B*3 lie on the surface.

On the other hand, imagine a line to be stretched from *M*, the middle point of *A*,*B*, to a point 3′, two paces round the curve at the top. If the line were straight, it would pass, not along the surface of the transformer, but actually up the inside, as may be seen both in the plan at (*b*) and in the perspective view at (*a*). If the line passed along the surface of the transformer it would be considerably curved towards the top, as may also be seen in those two views. In short, a piece of cord stretched from *M* to 3′ could not possibly lie in a straight line on the surface of the transformer. A proper realization of this will help the reader to understand the importance of correctly triangulating the surface of the object to be developed.

PATTERNS DRAWN TO SCALE

Sheet metal workers are often faced with the problem of developing a pattern much too large to be accommodated on one sheet of

Pattern enlarged from half scale to full size.

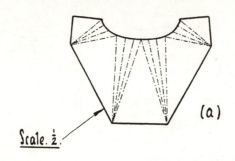

(a)

Scale. ½.

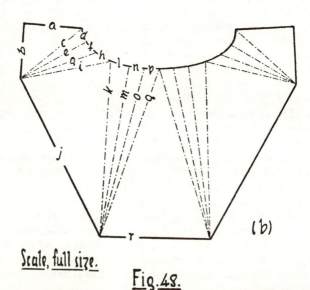

(b)

Scale, full size.

Fig. 48.

metal; such, for instance, as that of a hood perhaps 6, 9, or 12 ft. across its base. To lay down a full-sized plan and elevation for the purpose of developing a full-sized pattern would be a task of much difficulty, and would certainly prove a considerable time-waster. Patterns of large dimensions should be developed to scale, and then enlarged to full size when being transferred to the sheets. The method of drawing to scale also enables the craftsman to place the patterns to the best advantage for cutting them from the sheets, thus ensuring a minimum of waste by scrap.

There are two or three methods of enlarging the pattern from the scale drawing to the full size. Perhaps the simplest is, first, to mark off a straight line a little longer than the longest one needed in the full-sized pattern, and then, taking each line in order from the scale drawing, space it off with the compass along the line a number of times, equal to the scale of the drawing. Thus, if the drawing is one-quarter scale, each line should be spaced off four times. The result will give the full length of each line. Another good method is to carefully measure the length of each line, and multiply it by four, or by whatever number the denominator of the scale fraction might be, and then, using a tape measure as a trammel, mark off each line in the pattern. The best method is to use a scale rule, and read off the full length of each line directly, and then mark it off in the pattern. However, the use of the scale rule in this work will be dealt with more fully later in the course.

Fig. 48 shows a pattern at (a) developed by triangulation to a scale of one-half, and the same pattern at (b) enlarged to full size. This example will serve to illustrate how any pattern, either to scale or full size, may be transferred in this way. Each triangle is dealt with in order, and the pattern steadily developed. Beginning with the line marked a, which is made twice the length of the corresponding line in the half-scale pattern, the first triangle will be completed when the lines b and c are marked off, each twice the length of the corresponding lines in the smaller pattern. The second triangle will be completed when the lines d and e are dealt with, and the third triangle when f and g are added, and so on.

VARIETY OF TREATMENT

One of the difficulties which beset the beginner is to visualize the required form of the body, and to divide it into triangles correctly. A body incorrectly triangulated will produce an incorrect pattern, even though the golden rule of "plan length at right angles to vertical height" be carefully followed. One of the advantages of triangulation is its remarkable adaptability in application, particularly in the advanced stages of the work. In many cases the same problem may be dealt with in several different ways, according to the requirements of design and the ingenuity of the pattern drafter.

The illustrations shown at Fig. 49 represent four different designs of branch pieces to fit exactly the same positions. The relative merits of each, from the standpoint of efficiency, are considerations which

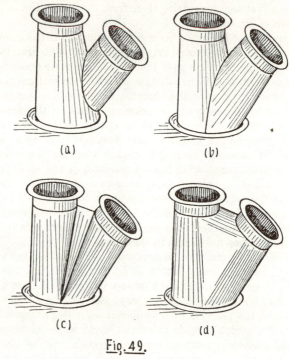

(a)   (b)

(c)   (d)

Fig. 49.

should be studied under the separate subject of Air Duct Design. It is sufficient for the moment that they illustrate the variety of treatment which the same conditions often afford.

TRANSFORMERS BETWEEN TWO PARALLEL PLANES

The broad principles of triangulation may be explained in a few words, but skill in the application of those principles cannot be gained in a few moments. The field of application is so broad and the variety of problems so great, that fresh posers are continually being encountered. However, in the early stages, the principles can best be grasped and understood by applying them to a few familiar problems, which, nevertheless, might well be turned round and viewed from slightly different angles. The problems dealt with in the First Course of triangulation are transformers between two parallel planes. The planes, for convenience, will be taken as hori-

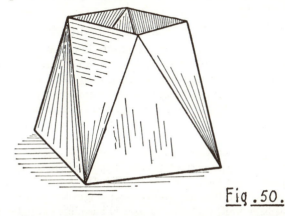

Fig. 50.

zontal. Therefore the top and bottom edges of the transformers will be level or horizontal. The vertical height will be the perpendicular distance between the two planes, and will be represented by a line

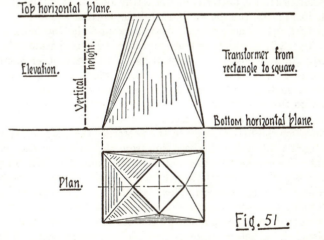

Top horizontal plane.

Elevation.

Vertical height.

Transformer from rectangle to square.

Bottom horizontal plane.

Plan.

Fig. 51.

drawn at right angles to the base from the top of the body. (See Fig. 51.) It will be observed that, by these conditions, every line drawn on the surface of the body from the top to the bottom will have the same vertical height. This, in the early stages, has the advantage of simplifying the work, since only one vertical height is needed in the elevation to serve for all the lines on the surface of the transformers. The plan will give all other details necessary for obtaining the pattern.

TWISTED SQUARE TRANSFORMER

One of the simplest and best examples to illustrate the principles of triangulation is afforded by the square-to-square transformer, in which the square at one end is placed diagonally to that at the other. In practice this occurs as a twisted connecting piece between two square pipes or portions of a duct, as shown in Figs. 50 and 53. Its surface is made up of flat triangles, with corners bent along the edges.

To develop the pattern, first draw the plan, showing the two squares placed in the correct positions in relation to each other, as in the plan above the pattern at (a), Fig. 52. A full elevation may be drawn if desired, but only the vertical height line will be needed, and should be set off, with the horizontal base line extended to any convenient length, as shown in the elevation at (a), Fig. 52. The corner points in the plan should now be numbered, beginning at the seam, as at 1,2,3,4,5,6,7,8,9. As already stated, every line forming the triangles in the pattern must be a true length. It will be observed that, in Fig. 52, the distances round the bottom square, 1,3; 3,5; 5,7; 7,9; 9,1; and also the distances round the top square, 2,4; 4,6; 6,8; 8,2; are already true lengths in the plan, because they are horizontal, and possess no vertical height. On the other hand, the distances 1,2; 2,3; 3,4; . . . 8,9; 9,2; which pass up and down between the top and bottom, do not represent true lengths. In order to obtain the true lengths of these lines, the distances from the plan must be triangulated against the vertical height.

Thus, for the first triangle in the pattern, take 1,2 from the plan and mark it off along the base line from the foot B of the vertical height. Then take the diagonal from the point 2 on the base line to the top T of the vertical height, and mark off this distance on the line 1′,2′ in the pattern. Next take 2,3 from the plan and mark it off along the base line at right angles to the vertical height; take the diagonal from the point 3 on the base line to the top T of the vertical height, and from the point 2′ in the pattern swing off an arc through the point 3′. Now take the true length line 1,3 direct from the plan, and from the point 1′ in the pattern describe an arc cutting the previous arc in 3′. Join 1′,3′ and 2′,3′. For the second triangle, place the plan length 3,4 at right angles to the vertical height, take the diagonal, and from point 3′ in the pattern, swing off an arc through the point 4′. Take the true length 2,4 from the plan, and describe an arc cutting the previous arc in point 4′. Join 2′,4′ and 3′,4′.

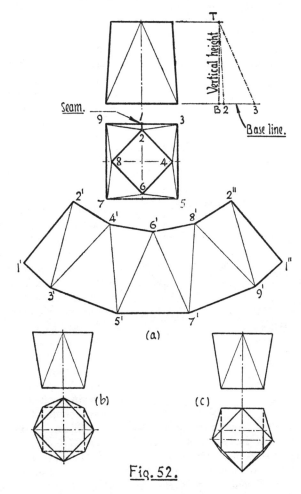

Fig. 52.

For the third triangle, repeat this process with lines 4,5 and 3,5. For the fourth, and remaining triangles, the process is repeated until the final triangle is completed.

If will be seen that the plan lengths of 2,3; 3,4; . . . 8,9; 9,2; are all equal. Therefore the distance $B,3$ on the base line serves for

all of them. The true lengths also are all the same. This condition
occurs when the top square is placed concentrically over the bottom,
but in a case such as that shown at (c), Fig. 52, where the top square
is off-centre with the bottom square, the plan lengths will vary, as
also will the true length. The example shown at (b), Fig. 52, represents the top and bottom squares of the same size. The development
of this transformer is carried out in a similar manner to that of (a),
although it is interesting to note that the pattern is actually a simple
rectangle, having a length
equal to the perimeter of one
of the squares.

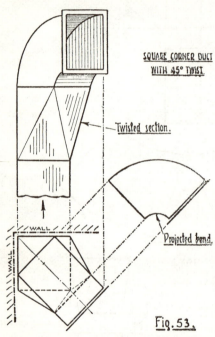

SQUARE CORNER DUCT
WITH 45° TWIST

Twisted section.

Projected bend.

Fig. 53.

The examples at (b) and
(c) are given as additional
exercises.

A SQUARE CORNER BEND

Whenever a square pipe
or duct undergoes a change
of direction at an angle
which involves a twist on its
axis, the type of transformer
dealt with in this, and the
foregoing, example is the
fulcrum on which the change
is effected. In this example,
shown in plan and elevation
in Fig. 53, a square duct is
seen passing up a wall
corner. A right-angle bend,
shown in projection, is required to come off at an angle
of 45 degrees to the normal axis of the duct. This necessitates that
the two sides of the twisted section facing the walls should be vertical.
The twisted section, or transforming piece, is shown in Fig. 54, and
the vertical sides referred to at 7,9 and 9,3. The top square 2,4,6,8
is therefore off-centre, and turned through an angle of 45 degrees
in relation to the bottom square 3,5,7,9. The seam is made from the
middle point 1 of the bottom side 9,3, up to the corner point 2 of the
top square.

To develop the pattern, set off the vertical height line from the
elevation. For the first triangle, take the plan length 1,2 and mark it

off from B along the base line at right angles to the vertical height. The diagonal 2,T will be the true length of this line, which should now be marked off in the pattern, as at 1',2'. Next, take the plan length 2,3 and mark it off from B along the base line. The diagonal 3,T will give the true length of this line. Take this distance 3,T in the compasses, and from 2' in the pattern swing off an arc through the point 3'. Now take the true length 1,3 from the plan, and from 1' in the pattern describe an arc cutting the previous arc in point 3'. For the second triangle, take the plan length 3,4 and mark it off from B along the base line. Take the diagonal true length 4,T, and from point 3' in the pattern swing off an arc through the point 4'. Next, take the true length 2,4 from the plan, and from 2' in the pattern describe an arc, cutting the previous arc in point 4'. For the third triangle, repeat the process with lines 4,5 and 3,5. For the remaining triangles the process is the same throughout.

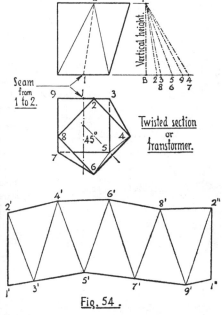

Twisted section or transformer.

Fig. 54.

If should be remembered that only those lines which pass from top to bottom or bottom to top should be triangulated against the vertical height. Those lines which form the perimeters of the squares are in horizontal planes and can therefore be taken as true lengths direct from the plan.

TWISTED RECTANGULAR DUCT

Sometimes it is required to twist a rectangular duct through 90 degrees to bring the broad side at right angles to its previous position. There are three typical cases of this problem worthy of note. First, there is the case in which the duct is turned concentrically on its axis, producing a connecting piece similar to that shown at (c), Fig. 55. In this case the pattern is formed of four similar trapeziums in alternate reverse positions, each representing one side of the transformer, and all together lie between two parallel lines. The second case is one in which the duct is turned off-centre in one direction only, so that one side may lie flat against a wall or ceiling. An example of this type is shown at (d), Fig. 55. In the third case the duct is turned off-centre in two directions, so that two sides may lie flat against two walls forming a wall corner, or, alternatively, against a wall and a ceiling according as the duct is vertical or horizontal.

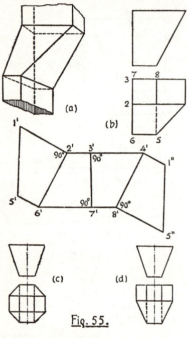

Fig. 55.

An example of the third class is shown perspectively at (a), and in plan and elevation at (b), Fig. 55.

The pattern may be developed by triangulation, but it will be simpler to lay out the sides direct by the use of the set-square. In the plan, the top rectangle is numbered 1,2,3,4, and the bottom rectangle 5,6,7,8. It will be observed that the corner from point 3 runs vertically downwards to point 7. Let this corner constitute a bend in the middle of the pattern; then the two sides forming the bend may readily be set out. First draw the bend line 3',7', and set off 2',4' and 6',8' at right angles to it. Mark off 3',2'; 3',4'; and 7',6'; 7',8' equal to the corresponding distances from the plan. Join 2',6' and 4',8'. For the other two sides, it will be seen from the plan that the top edge 2,1 is at right angles to the corner 2,6. Therefore the distance 2,1 from the plan can now be set off in the pattern at right angles to 2',6', and the distance 6,5 from the plan set off from 6' parallel to 2',1'. Join 1',5'. Similarly, in the plan it will be seen that the bottom edge 8,5 is at right angles to the corner 8,4. Therefore the distance 8,5 from the plan can be set off in the pattern at right angles to 8',4', and the distance 4,1 from the plan set off from 4' parallel to 8',5''. Join 1'',5''.

The examples at (c) and (d) are left as additional exercises.

THE SQUARE-TO-CIRCLE TRANSFORMER

A problem of very common occurrence in sheet metal work, particularly in pipe and duct work, is that of the square-to-circle transformer, often called a tallboy. Its object in duct work is to transform a square or rectangular pipe to a round pipe, or to connect a round

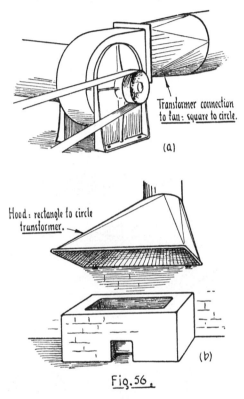

Transformer connection to fan: square to circle.

(a)

Hood: rectangle to circle transformer.

(b)

Fig. 56.

pipe to a square or rectangular hole, such as the outlet of a centrifugal fan. This type of transformer also takes the form of hoods over hearths and furnaces to collect the fumes which rise up through the pipe at the top. (See Fig. 56 (a) and (b).) In general practice it is encountered in a variety of ways, almost too numerous to mention.

The simplest example of this class is such as that shown in Fig. 57, in which the centre of the circle in the plan coincides with that of the square, and in which the diameter of the circle is smaller than the width of the square. The method of developing the pattern, however, is the same for every case, whether the circle is the same size as or larger than the square, or whether the circle is off-centre one way or both ways with that of the square.

To develop the pattern, divide the circle in the plan into twelve equal parts. Any other number would do as well, but twelve is very

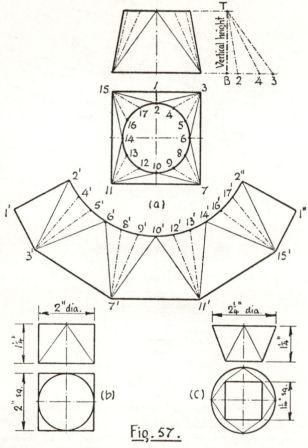

Fig. 57.

convenient, as the divisions can be obtained with the compasses without altering the radius. Assuming the seam to be up the middle of one side as at 1,2, Fig. 57, number the points, beginning at the seam, as shown at 1,2,3,4 . . . 14,15,16,17. Project a vertical height line from the elevation, and extend the base line sufficiently to accommodate the longest plan length.

For the first triangle, take the plan length 1,2 in the compasses, and mark it off from B along the base line at right angles to the vertical height. Take the true length diagonal from 2 to the top T of the vertical height, and set off $1',2'$, in the pattern. It will be observed that this first line in the pattern may be set off anywhere and in any position, since the rest of the pattern will follow accordingly. However, a little care and foresight are usually needed to place the first line so that the pattern following up will not run off the sheet or the paper. Next take 2,3 from the plan, mark it off from B along the base line at right angles to the vertical height, take the true length diagonal from 3 to the top T, and from point $2'$ in the pattern swing off an arc through point $3'$. Next take the true distance 1,3 from the plan, and from the point $1'$ in the pattern describe an arc cutting the previous arc in point $3'$. Join $2',3'$ and $1',3'$.

For the second triangle, take 3,4 from the plan, mark it off along the base line, take the true length diagonal, and from point $3'$ in the pattern swing off an arc through the point $4'$. Next take the true length 2,4 direct from the plan, and from point $2'$ in the pattern describe an arc cutting the previous arc in point $4'$. Join $3',4'$. For the third triangle, repeat this process with the plan lengths 3,5 and 4,5; and again for the fourth triangle with plan lengths 3,6 and 5,6. For the fifth triangle, repeat the process with plan lengths 6,7 and 3,7; but observe in this case that the triangle is reversed in position. The remainder of the pattern should now be quite easy to follow, since it is a repetition of these processes right through. The line from $2'$ to $2''$ should be a curve, and not a series of short straight lines.

METHOD OF NUMBERING THE POINTS

One of the best aids to clear development in triangulation is a satisfactory method of numbering the points of division in the plan and elevation. Many different methods are in use, but the author prefers one in which the consecutive numbers 1,2,3,4 . . . and so on, can be used right round the body from seam to seam. In many cases of triangulation, the arrangement of the lines forming the triangles produces a continuous zigzag line on the surface between the top and bottom edges of the body, as shown at (a), Fig. 58. Examples of this class were given in the previous section, on square-to-square transformers, and many other examples are to follow. The zigzag line need not be regular in form, as at (a), Fig. 58, but may take almost any shape, such as the example at (b). The chief point to observe is that, beginning at the seam with 1 to 2, the consecutive numbers are placed alternately at top and bottom as the zigzag line forming the triangle passes round the body. This is a very simple method of numbering, and has the advantage that, if the work of development be left for a time, it can be picked up with confidence at the exact spot where it was left off.

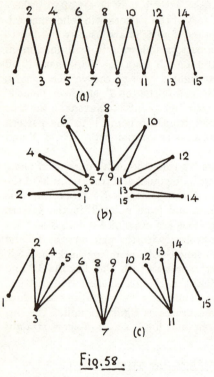

Square-to-circle transformers introduce a little variation on this arrangement, the principle of which is shown at (c), Fig. 58. From this diagram it will be seen that the continuous zigzag line is formed by 1,2,3,6,7,10,11,14,15. But there are other lines radiating from points 3,7, and 11, as 3,4 and 3,5. The method of procedure, in a case like this, is to begin at 1 and follow up in zigzag form with 2, 3, and 4. From point 4, there is no return to the base, other than back to 3. Retrace back to 3, and proceed to point 5. Again retrace back to 3, and proceed to point 6.

Fig. 58.

It is now possible to get back to the base from 6 to 7. From point 7 the process is repeated as from point 3, but this time with 7,8; 7,9; and 7,10; and then back to 11. This is repeated again from point 11.

It will be found that this method of numbering can be applied to all cases, and will be of considerable advantage when complicated problems are dealt with.

OFF-CENTRE RECTANGLE-TO-CIRCLE TRANSFORMER

To develop the pattern for a rectangle-to-circle transformer, off-centre both ways, set out the plan and elevation to suit the required dimensions, as shown at (a), Fig. 59. Assuming that the seam is to be on the triangle at the shorter end, divide the surface into triangles and, beginning at the seam, number the points as shown in the diagram.

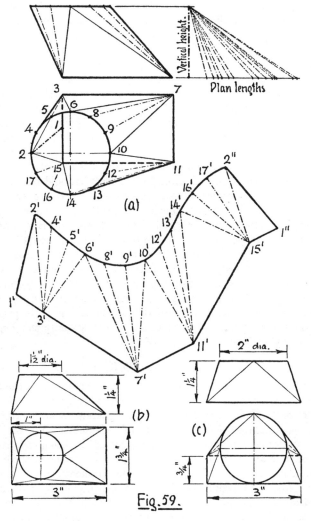

Fig. 59.

For the first triangle, take the plan length 1,2, and mark off at right angles to the vertical height; take the true length diagonal, and set off 1',2' in the pattern. Next take the plan length 2,3, and mark off at right angles to the vertical height; take the true length diagonal, and from 2' in the pattern swing off an arc through point 3'. Take the true length 1,3 direct from the plan, and from point 1' in the pattern, describe an arc cutting the previous arc in point 3'. Join 1',3' and 2',3'. For the second triangle, take the plan length 3,4, and mark it off at right angles to the vertical height; take the true length diagonal, and from point 3' in the pattern swing off an arc through point 4'. Next take the true length 2,4, direct from the plan, and from point 2' in the pattern, describe an arc cutting the previous arc in point 4'. Join 3',4'. For the third triangle, repeat this process with plan lengths 3,5 and 4,5; and again for the fourth triangle with plan lengths 3,6 and 5,6. For the fifth triangle repeat the process with plan lengths 6,7 and 3,7; but, again, observe that the triangle is reversed in position. The remainder of the pattern should now be quite easy to follow, since it is a repetition of these processes right through. The line from 2' to 2" should be a curve.

It will probably be noticed that this description is almost a repetition of that for the development of the problem shown at Fig. 57. This fact is worthy of particular observation, since it brings out one or two important points. All square-to-circle transformers between two parallel planes, which have been numbered in accordance with this rule, have precisely the same method of development. However different the problems might appear in plan and elevation, the method of procedure is exactly the same. It is only a question of carefully following the routine of development. A little practice will show the simplicity of the method and how easily this type of problem can be dealt with.

THE GROUPING OF SIMILARITIES

The art of pattern-drafting is intimately associated with the craft of sheet metal work. Before the application of geometry became popular, patterns used to be drafted by guesswork and numerous devices accumulated over long years of experience. Yet methods of trial and error can never be accurate, except by the lavish expenditure of time and trouble. Now, since accuracy is one of the essential conditions of modern production, the application of geometry to the solution of problems of pattern-drafting is an economical method of ensuring success. The principles of surface development are most profitably studied by taking problems in systematic order rather than by casual selection. The grouping of similarities is a valuable help in bringing out the fundamental principles. The term "similarities" here refers to similarities of methods of development rather than similarities of form.

In this first course of triangulation, the problems dealt with are transformers which lie between two parallel planes. Up to this point have been those which transform from a square at one end to a similar square placed diagonally at the other, and these were followed by the tallboy type which transforms from a square at one end to a circle at the other. Those which now follow in progressive order are such as transform from an oval or other curved outline at one end to a circle at the other; those which transform from a semicircle at one end to a circle at the other; and those which transform from a combined rectangle and semicircle to a circle at the other. In practice, these problems find application in many forms of hoods and hoppers. The illustration in Fig. 60 shows, in elevation and plan, a connecting piece which fits at the base on one half of the top end of a cylindrical body, and transforms to a circular pipe above. The development of the pattern for this type is shown in Fig. 62.

THE OVAL-TO-CIRCLE TRANS-FORMER

To develop the pattern for the

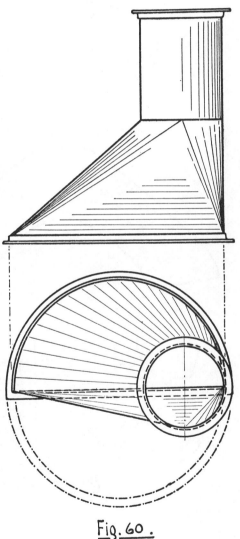

Fig. 60.

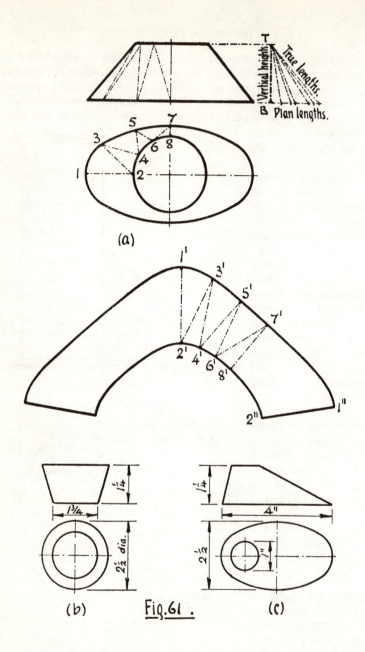

T

True lengths

Vertical height

B

Plan lengths.

5   7
3   6 8
    4
1   2

(a)

1'
        3'
            5'
                7'
2' 4 6' 8'
2"   1"

1¼

1¾

2½ dia.

(b)

1¼

4"

2½

1"

(c)

Fig. 61.

oval-to-circle transformer, first set out the plan and elevation as shown at (a) Fig. 61. In the plan, divide one quarter of the circle into three equal parts and the corresponding quarter of the oval similarly. One quarter will be sufficient since the plan is symmetrical about both axes. Number the points from 1 to 8. A zigzag line drawn between these points will divide that part of the surface into triangles. Project a vertical height line in the elevation, and extend the base line along which to mark off the plan lengths.

For the first triangle, take the plan length 1,2, and mark it off from B at right angles to the vertical height. Take the diagonal true length line and set off 1',2' in the pattern. Next take the plan length, 2,3, and mark it off at right angles to the vertical height; take the diagonal true length, and from 2' in the pattern swing off an arc through point 3'. Take the true length 1,3 direct from the plan, and from point 1' in the pattern describe an arc cutting the previous arc in point 3'. Join 1',3' and 2',3'.

For the second triangle, take the plan length 3,4, and mark it off at right angles to the vertical height; take the diagonal true length, and from point 3' in the pattern, swing off an arc through point 4'. Next take the true length 2,4, direct from the plan, and from point 2' in the pattern describe an arc cutting the previous arc in point 4'. Join 3',4' and 2',4'. For the third triangle, repeat this process with plan lengths 3,5 and 4,5; and again for the fourth triangle with plan lengths 4,6 and 5,6. The remaining two triangles which complete the quarter pattern are similarly obtained from the plan lengths. 5,7; 6,7; and 6,8; 7,8. The full pattern may now be completed by repeating or duplicating this quarter in the order shown in the figure. The examples at (b) and (c) are given as additional exercises.

THE SEMICIRCLE-TO-CIRCLE TRANSFORMER

Although the semicircle-to-circle transformer appears to be quite different from the previous example, the method of procedure in developing the pattern is precisely the same. A little more care, however, will be needed in following the numbered points. The seam in this case (Fig. 62) extends from point 1 on the middle of the straight side to point 2 on the corresponding quarter point on the circle above. It will be seen that the quadrant in the circle from 2 to 6 is connected to the corner point 3 by the lines 3,2; 3,4; 3,5; 3,6; and that the quadrant from 2 to 18 is connected to the other corner point 17 by the lines 17,18; 17,19; 17,20; 17,2. A certain amount of overlapping makes it difficult to show these lines clearly.

The line in the plan, for instance, from 17 to 18, lies vertically upwards, and is, therefore, represented by a single point which takes both numbers. Also, the line from 17 to 19 lies so close to the curve

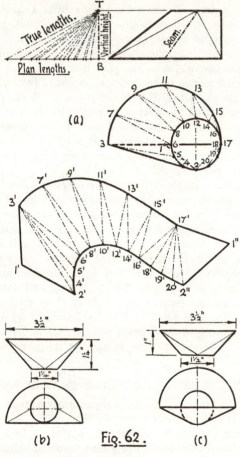

(a)

(b)   Fig. 62.   (c)

of the circle that it cannot be clearly separated. Again, the line from 2 to 3 almost coincides with that from 3 to 4. Nevertheless, with careful attention to these points it should not be difficult to follow the sequence of numbers.

To develop the pattern, set out the plan and elevation to the required dimensions, and, in the plan, divide the top circle into twelve equal parts and the bottom semicircle into six equal parts. Beginning at the seam, number the points from 1 to 20 as shown in the diagram. It will be noted in this case that the whole of the surface must be triangulated for development, as it is not symmetrical about any axis. Project a vertical height line in the elevation, and extend the base line, along which to mark off the plan lengths. It will also be noted that the vertical height and the base line are projected on the left-hand side instead of the right. This, however, makes no difference at all to the method of development.

For the first triangle, take the plan length 1,2, and mark it off from B at right angles to the vertical height. Take the diagonal true length and set off 1′,2′ in the pattern. Next take the plan length 2,3, and mark it off at right angles to the vertical height; take the diagonal true length, and from 2′ in the pattern swing off an arc through point 3′. Take the true length 1,3 direct from the plan, and from point 1′ in the pattern describe an arc cutting the previous arc in point 3′. Join 1′,3′ and 2′,3′. For the second triangle, take the plan length 3,4 and mark it off at right angles to the vertical height; take the diagonal true length, and from point 3′ in the pattern, swing off an arc through point 4′. Next take the true length 2,4, direct from the plan, and from point 2′ in the pattern, describe an arc cutting the previous arc in point 4′. Join 3′,4′ and 2′,4′. For the third triangle, repeat this process with plan lengths 3,5 and 4,5, and again for the fourth triangle with plan lengths 3,6 and 5,6. For the remainder of the pattern the process is further repeated using the plan lengths in the order shown in the diagram.

It should be remembered that in all cases of transformers between horizontal planes, the distances taken round the top and bottom perimeters are already true lengths in the plan, and do not require to be triangulated against the vertical height. On the other hand, all those distances taken from the plan which pass up and down from bottom to top and top to bottom, do not represent true lengths, and must be triangulated against the vertical height, when the diagonal will give the required true length.

THE RECTANGLE, SEMICIRCLE-TO-CIRCLE TRANSFORMER

The example shown at (a), Fig. 63, is a combination of half a cone and half a tallboy.

It is a typical form of transformer which serves as a hopper with its back to a wall. The circular hole at the bottom may be proportionately large or small, or off-centre according to circumstances. The method of development, however, is exactly the same in each case. It is merely a question of carefully triangulating the surface in the manner shown at (a) and following the course of directions for the pattern given in the previous example. Since the directions for

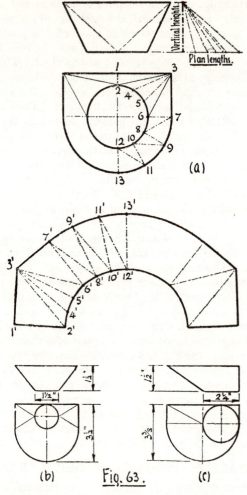

Fig. 63.

developing the pattern are similar to those of the example in Fig. 62, these at (a), (b), and (c), Fig. 63, are left as additional exercises.

FUNDAMENTAL PRINCIPLES

Pattern-developing is a subject which may be approached from more than one standpoint. It is a comparatively easy task to follow implicitly a fully detailed draft of directions for one particular problem. The completion of one may be followed by the solution of several more problems, and a certain sense of satisfaction may be experienced as one after another is added to the list. Yet, when each problem becomes a formula peculiar to itself, which must be memorized in order to dispense with the directions, the student cannot progress very far. The subject soon becomes too great a tax on the memory. The best and most satisfactory method of tackling the subject is to concentrate on the fundamental principles, and digest them by practising on suitably arranged problems. This may not be so easy in the beginning, but the persevering student will eventually emerge with a more lasting and reliable knowledge of pattern development which he can apply with confidence to his own range of problems without feeling the need of ready-made recipes for such cases as depart from the usual order. This course of geometry is designed to develop these principles, and to provide suitable examples for practice, as this, like many other studies, needs practice in order to gain facility in the application of the principles.

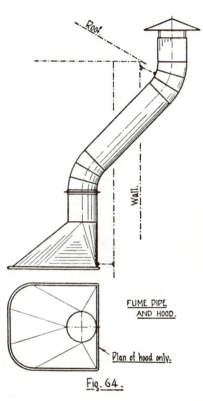

FUME PIPE
AND HOOD.

Plan of hood only.

Fig. 64.

Fig. 64 shows a typical form of hood over a hearth, with the fume pipe taking an easy direction through the wall, such that the upward draught meets with a minimum of resistance. Sometimes, when there is much smoke, it is necessary to arrange for cleaning by fitting stumps with easily removable caps. The hood is a little different from the ordinary tallboy, inasmuch as the two front corners at the bottom are quadrants with a good radius. The development of the pattern is shown at Fig. 65.

HOOD WITH CORNER RADII

To develop the pattern for the hood shown in Fig. 65, set out the plan and elevation to the required dimensions; divide the surface into triangles and number the points in the manner shown in the

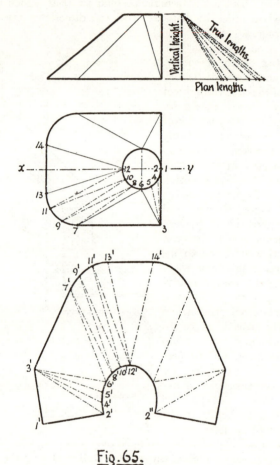

Fig. 65.

diagram. Since the plan is symmetrical about the centreline xy, only half of the figure need be triangulated. However, it will be noted that the point 14 is carried over to that opposite point 13, so that the full triangle 12,13,14 may be developed in one. The remaining part of the pattern is then a reversed repetition of that up to point 13.

For the first triangle, take the plan length 1,2, and mark it off at right angles to the vertical height; take the diagonal true length and set off 1′,2′ in the pattern. It will be seen in this case that the plan length 1,2 is a single point, and has no actual length. The vertical height, therefore, becomes the true length. Next take the plan length 2,3, and mark it off at right angles to the vertical height; take the diagonal true length, and from 2′ in the pattern swing off an arc through point 3′. Take the true length 1,3 direct from the plan, and from point 1′ in the pattern describe an arc cutting the previous arc in point 3′. Join 1′,3′ and 2′,3′. For the second triangle, take the plan length 3,4, and mark it off at right angles to the vertical height; take the diagonal true length, and from point 3′ in the pattern, swing off an arc through point 4′. Next take the true length 2,4 direct from the plan, and from point 2′ in the pattern describe an arc cutting the previous arc in point 4′. Join 2′,4′ and 3′,4′. For the third triangle, repeat this process with plan lengths 3,5 and 4,5; and again for the fourth triangle with plan lengths 3,6 and 5,6. For the fifth triangle repeat the process with plan lengths 6,7 and 3,7; but observe in this case that the triangle is reversed in position. The remainder of the pattern should now be quite easy to follow, since it is a repetition of these processes right through. The line from 2′ to 2″ should be a curve, and not a series of short straight lines.

PLANT WORK

The illustration in Fig. 66 shows, in front elevation and side elevation, an arrangement of plant for conveying material to a grading machine from a position on a floor above. The horizontal distance may be anything from a few feet to 100 or more; the length of the chain conveyer may be adjusted to suit. The dimensioned parts are for such work as would be required of the sheet metal worker, although the chain conveyer casing, which is not dimensioned, may also be of sheet metal. This arrangement is only one of innumerable scores where sheet metal work is required in the installation of plant equipment. Work of this kind calls for skill and resourcefulness, and involves a great deal of pattern development. The examples here given are similar to some which have already appeared in this course.

The hopper, illustrated at (a), intended to receive the material, is, geometrically, the frustum of a right cone. The pattern should therefore be developed by the Radial Line Method. A similar example to this, in the form of a conical jug top, was given as an additional exercise in Fig. 9. The chute at (b) delivers the material

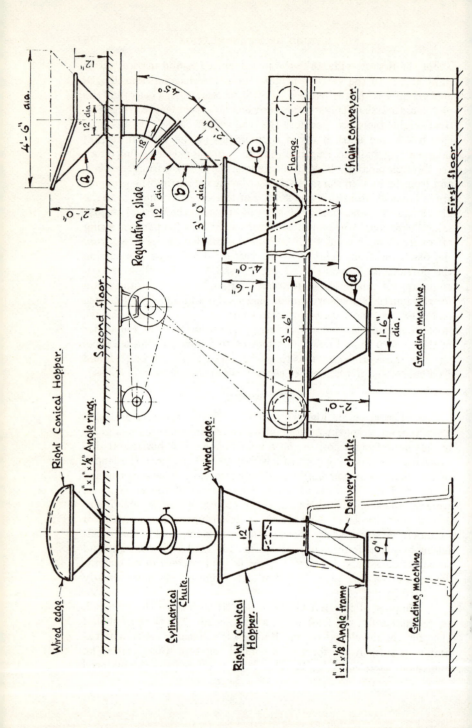

Right Conical Hopper.

Wired edge.

1" x 1" x ⅛" Angle rings.

Second floor.

Cylindrical Chute.

Wired edge.

Right Conical Hopper.

12"

Delivery chute.

9"

1" x 1" x ⅛" Angle frame.

Grading machine.

4'- 6" dia.

12"

12" dia.

45°

2'- 0"

18"

Regulating slide 12" dia.

12" dia.

ⓐ

ⓑ

ⓒ

3'- 0" dia.

Flange.

Chain conveyor.

4'- 0"

1'- 6"

3'- 6"

ⓓ

1'- 6" dia.

2'- 0"

Grading machine.

First floor.

to the hopper below. This chute is an example involving the Parallel Line Method of development, and examples of the lobster-back bend portion will be found in Figs. 35 and 36. The lower part of the chute is similar to the portion of a pipe elbow as shown in Fig. 22. The hopper on the chain conveyer at (c), which receives the material from the chute, is of right conical form, and fits over the conveyer casing to deliver the material to the bottom chain, as will be seen in the side elevation. There has been no previous example quite like this, but no difficulty should be experienced in developing the pattern if the principles of the Radial Line Method have been carefully followed. The chute at (d), which delivers the material from the conveyer to the grading machine, is a rectangle to circle transformer. An example similar to this was given as an additional exercise at (c), Fig. 59. These examples in Fig. 66 may be regarded as additional exercises.

Air-duct systems form another type of plant work which embraces a large variety of geometrical problems. Hoods, transformers, branch connections, and junction pieces are among the details which offer much scope for ingenious design and skilful execution. In air-duct work efficiency should not be sacrificed for simplicity. Simplicity is an excellent maxim where efficiency is not impaired by its adoption. However, the problems of efficiency in connection with air-duct design need special consideration.

SECOND COURSE

CHAPTER 5

THE RADIAL LINE METHOD

In planning the Second Course of geometry for sheet metal work, many fresh principles and problems have to be considered, yet the main classification of methods of development remains the same. The Radial Line Method, the Parallel Line Method, and Triangulation are the only fundamental methods on which the art of pattern-drafting is based, but there are numerous subsidiary principles which govern the minor details of procedure in solving any particular type of problem. For instance, problems of both the right cone and the oblique cone are solved by the Radial Line Method, yet there are characteristic differences which make the two types distinct.

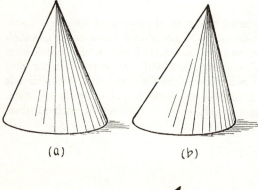

(a)   (b)

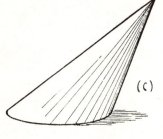

(c)

Fig. 67.

The Second Course is based on the following arrangement—

(a) Developments by the Radial Line Method of patterns involving problems of the oblique cone and its frustums.

(b) Developments by the Parallel Line Method of patterns for branch joints between pipes of unequal diameters. Further examples of moulding, such as roof finials and downspout heads.

(c) Intersections of the right cone and cylinder around a common central sphere.

(d) Developments by the Method of Triangulation of patterns for transformers between planes inclined at an angle to each other.

THE OBLIQUE CONE

The oblique cone is so important in pattern-drafting that a few of its outstanding properties are well worth careful observation before turning to the development of its surface. To compare it with the right cone—

A RIGHT CONE has a circular base, and its apex lies perpendicularly over the centre of the base.

An OBLIQUE CONE has a circular base, but its apex does not lie perpendicularly over the centre of the base. The axis of the oblique cone, therefore, leans to one side of the perpendicular.

Fig. 67 shows a right cone at (a) and oblique cones at (b) and (c). An oblique cone cut by any plane parallel to the base presents a circle at the plane of cutting. Thus, referring to Fig. 68, the oblique cone at ABC is cut by a plane DE parallel to the base BC. The shape of the cone at DE is, therefore, a circle.

THE SUBCONTRARY SECTION

Like the oblique cylinder, the oblique cone possesses a subcontrary section, which presents a circle at a plane not parallel to the base. In Fig. 68 (a) the plane DE is parallel to the base and its cross-section is circular. The corresponding subcontrary section occurs at FG, which occupies a position such that, from the apex A, AG is equal to AE and AF is equal to AD. Also, the angle AGF is equal to the corresponding opposite angle AED, and the angle ADE is equal to the corresponding opposite angle AFG. By these conditions, FG equals DE, and the section at FG is a circle equal to the circle at DE. Any plane cutting the oblique cone parallel to FG or to DE presents a circular cross-section, but at no other angle can a circular cross-section be obtained.

One other cross-section is important. Any cutting plane through the cone at right angles to its axis, as at HC, such that the side AH equals AC, presents an ellipse at HC. The importance of a clear understanding of this will be seen by reference to Fig. 68 (b). The cone AHC appears to be a form of right cone with an elliptical base, whereas it is really an oblique cone cut through at right angles to its central axis. Circular cross-sections occur at DE and FG, or at any plane parallel. to either. It will be observed that the

geometrical properties of this cone are identical with those of the
top part *AHC* of the cone shown in Fig. 68 (*a*).

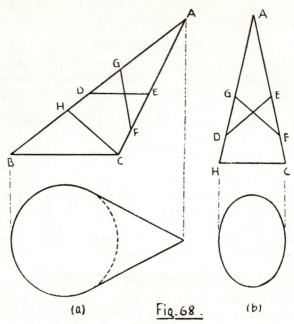

(a)   Fig. 68.   (b)

THE OBLIQUE CONE DEVELOPED

Whereas, in the case of the right cone, the distance from the apex
to the base is the same at all positions round the surface, the oblique
cone varies from point to point.

To develop the pattern for the oblique cone, draw the elevation
to the required dimensions, and describe a semicircle on the base.
Divide the semicircle into six equal parts and number the points as
from 1 to 7 in Fig. 69. From the apex *A*, drop a vertical line to meet
the base line produced in *A'*. A line from *A'* drawn as a tangent to
the semicircle will then complete a half-plan of the cone. With the
compasses at *A'*, which represents the apex in the plan, draw arcs
from points 2,3,4,5,6, to the base of the cone. Next, from the apex *A*
in the elevation, swing arcs into the pattern from the points thus
obtained on the base line, including the two outside points 1 and 7.
Take one of the divisions from the semicircle, as 1 to 2, or 2 to 3, and
beginning at any suitable point on the inner arc, as at 1', mark off
the points 1',2',3',4',5',6',7' . . . 1", by stepping over from one

line to the next. A curve drawn through these points from 1' to 1"
will give the base curve in the pattern. Join 1',A, and 1",A.

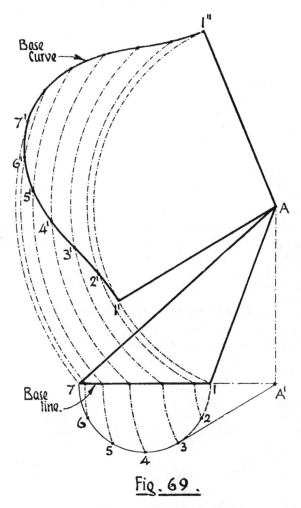

Base Curve

Base line.

Fig. 69.

THE OBLIQUE CONE FRUSTUM

It is very rare in practice that the full oblique cone is required.
It is nearly always truncated, and often cut by curved surfaces. The
diagram in Fig. 70 shows a similar cone to that of Fig. 69, but it is
truncated, or cut off at *MN*.

To develop the pattern for the frustum or bottom portion, proceed exactly as in the previous example to obtain the base curve. Then, for the top curve, draw lines from the points on the base line to the apex A. Where these lines cross the top line MN, a series of points similar to those on the base line will be obtained. From these points, with apex A as centre, swing another series of arcs into the pattern. Next, join the points $1',2',3',4',5',6',7', \ldots 1''$, on the base curve to the apex A. Where these lines cross the arcs from the line MN, a series of points will be obtained through which to draw the top curve as shown in Fig. 70. This curve will be similar to the base curve, but smaller.

An important observation might be made here. The golden rule of triangulation, in which the plan length at right angles to the vertical height gives a diagonal which is the true length, also forms the basis of the Radial Line Method. For example, the plan length from A' to 3 is swung round to $3''$ on the base line. Since A,A' is the vertical height of the cone and $A',3''$ is at right angles to it, it follows that the diagonal $A,3''$ is a true length line. This, in turn, is swung into the pattern at $A,3'$. Similarly, each of the other lines in the pattern radiating from the apex A is a true length.

The two smaller diagrams at the bottom in Fig. 70 are given as additional exercises.

APPLICATIONS OF THE OBLIQUE CONE

The oblique cone has many practical applications in sheet metal work. Among the important examples are those in which portions of the cone, in conjunction with flat sides, make up the body required. Fig. 71 shows a method of transforming a cylindrical duct to a flat one with semicircular sides in order to pass through a confined space. One section is shown in Fig. 72, and it will be seen that the portion $ABCD$ is really half an oblique cone, because the plane of the semicircle AD is parallel to the plane of the semicircle BC. The portion $AEFD$ is also a similar half of an oblique cone, and the two flat triangles ABE and DCF between them complete the body. The development of this type of transformer is shown in Fig. 74, which has a somewhat longer top, and is more akin to a fishtail nozzle.

A SMOKE HOOD

The development shown in Fig. 73 is of a smoke hood which is fixed, on the vertical side, direct to the wall of the chimney or flue. The top of the hood is blanked up, and the smoke passes through a

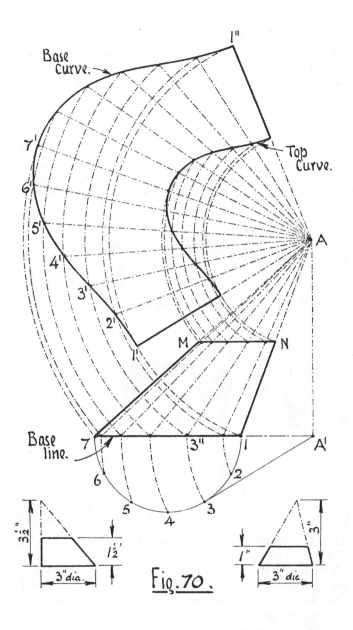

Base Curve.

Top Curve.

Base line.

A

M   N

A'

Fig. 70.

$3\frac{1}{2}''$ $1\frac{1}{2}''$ 3″dia.

1″   3″   3″dia.

hole in the chimney wall near the top. This type of hood is formed of half an oblique conic frustum with a flat piece on each side. The half-conic frustum is the semicircular portion on the left of *BIKF* in

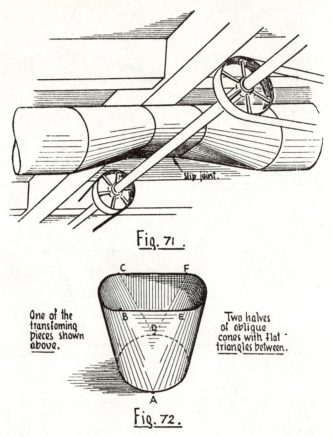

Fig. 71.

One of the transfoming pieces shown above.

Two halves of oblique cones with flat triangles between.

Fig. 72.

the plan, and the two flat sides are *ABJI* and *GFKH*. The apex of the conic portion is at *X*.

To develop the pattern, set out the plan and elevation to the required dimensions, as shown in Fig. 73. Join *BX* and *FX* in the plan. Divide the quadrant *BCDE* into three equal parts, and, with *X* as centre, swing arcs from *A,B,C* and *D* on to the centreline *EX*. Transfer these points on *EX* vertically upwards to the base line in the elevation. Letter the points on the base line with the corresponding letters in the plan, as at *A′,B′,C′,D′,E′*. With *X′* in the

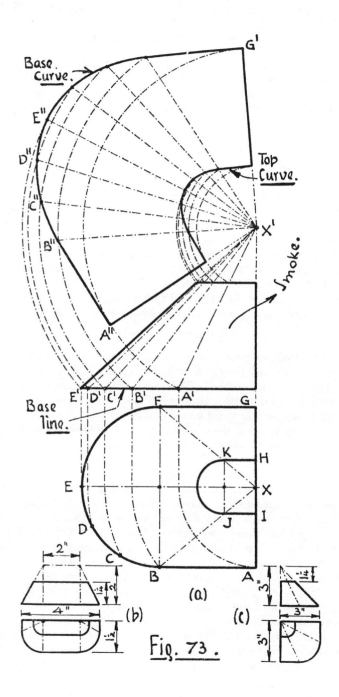

Base Curve.

E''

D''

C''

B''

Top Curve.

X'

A''

Smoke.

Base line.

E' D' C' B' F A' G'

K H

E X

D J I

C

B A

(a)

2"

½

1¼

4"

1½

(b)

(c)

3"

1¼

3"

3"

3"

Fig. 73.

elevation as centre, swing arcs from these points into the pattern. Next, in consecutive order, take the distances AB,BC,CD,DE from the plan, and, beginning at a convenient point on the arc from A', mark off these distances in the pattern, stepping over from one line to the next, as at A'',B'',C'',D'',E''. Repeat these markings in the reverse order to G'. A straight line drawn from A'' to B'', and a curve from B'' to E'', and these repeated in the reverse order to G', will give the base curve in the pattern. Join A'',X' and G',X'. To obtain the top curve in the pattern, join A',B',C',D',E' to the apex X' in the elevation. From the points where these lines cross the top of the hood, swing arcs into the pattern, using X' as centre. Join the points B'',C'',D'',E'' in the pattern to the apex X'. These lines, crossing the arcs from the top of the hood, will give the necessary points for the top curve, which should be similar to the base curve, but smaller. The two examples given at the foot of the figure are smoke hoods of a similar character. That shown at (b) is of double width to cover two hearths, and that shown at (c) is suitable for a corner position. These are left as additional exercises.

A FISHTAIL NOZZLE

Perhaps the simplest way to make a fishtail nozzle is to flatten the end of a pipe and trim it to suit, but given dimensions cannot be worked to by this method. The width of the fishtail can never be wider than $1\frac{1}{2}$ times the diameter of the pipe. Moreover, the area of the flattened end would be reduced in accordance with the degree of flattening, which in the case of an air duct, would be an undesirable condition. The nozzle shown at (a) Fig. 74, has semicircular ends at the top, and is composed of two half-oblique cones with flat triangles between them, similar to the transformer shown in Fig. 72.

To develop the pattern, first set out the plan and elevation to the required dimensions, and locate the apex of the half-oblique cone in both views, as at X and X', Fig. 74. Divide the quadrant $ABCD$ in the plan into three equal parts, and with X as centre, swing arcs from B,C and D on to the centre line XY. Transfer these points on XY, including A, vertically upwards to the base line in the elevation. Letter the points on the base line with the corresponding letters in the plan, as at A',B',C',D'. With X' in the elevation as centre, swing arcs from these points into the pattern. Next, in consecutive order, take the distances AB,BC,CD, from the plan, and, beginning at a convenient point on the arc from A', mark off these distances in the pattern, stepping over from one line to the next, as at A'',B'',C'',D. These will give points on the base curve, which may now be joined

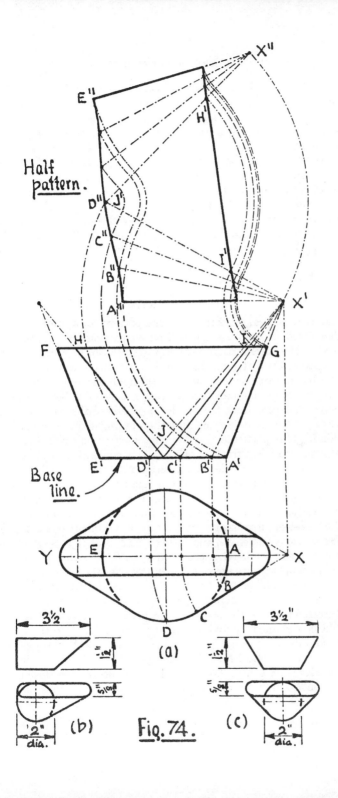

Half pattern.

Base line.

3½" 1½" 2" dia. (b)

(a)

D

C

B

A

Y E X

3½" 1½" 5/8" 2" dia. (c)

Fig. 74.

to the apex X'. Next, from B',C',D', on the base line, draw lines to the apex X', and from the points where these lines cross the top of the nozzle, swing arcs into the pattern, using X' as centre. Where these arcs cross the lines from A'',B'',C'',D'', points will be obtained through which to draw the top curve. The next step is to transfer the line $D''X'$, and all the points on it, to the position $D''X''$, by using D'' as centre, and swinging arcs from all the points on $D''X'$. To determine the position of the line $D''X''$, take IH in the compasses from the elevation, and mark off $I'H'$ in the pattern on the inner arc as shown in the diagram. From D'' draw through H' to meet the outer arc in X''. Join I',H'. This completes the flat triangle HIJ in the pattern. The remaining conical portion of the pattern is a reverse repetition of the first part $A''D''X'$ on the other side of the line $D''X''$.

A HIP BATH

There are several ways in which the pattern for a hip bath might be developed. A good deal depends on the particulars and conditions set out in the specification. The hip bath shown in Fig. 75 may be developed either by triangulation or by the radial line method. Some hip baths admit of development by triangulation only. In Fig. 75, a half-plan is placed above the elevation instead of below, in order to leave the space clear for the development of the pattern. This, however, should in no way confuse the working if that point is borne in mind. This particular bath is composed of half a frustum of a right cone and half a frustum of an oblique cone, with two flat pieces between them. In the elevation, because the plan on AB is a semi-circle and the plan on CD is also a semicircle, and the axis DAX is not at right angles to the planes of these semicircles, the portion $ABCD$ is that of the oblique cone frustum, with its apex at X. Again, in the elevation, because the plan on EG is a semicircle and the plan on HI is also a semicircle, and the axis EIX is at right angles to the planes of these semicircles, the portion $EFGHI$ is that of the right cone frustum, with its apex at X. In this case the plane at EF cuts the right cone obliquely. The two flat pieces are, one at $ADEI$ and the other at the corresponding position on the other side of the bath.

With reference to the inverted half-plan, the curve from E' to F' should be obtained by dividing the quadrant $E'G'$ into three equal parts and drawing radial lines from the plan apex X' through these points. Drop vertical lines from the points on $E'G'$ to the line EG in the elevation below. From the apex X, draw lines through the points on EG to meet the line EF. From the points thus obtained on EF, draw vertically upwards to meet the radial lines in c and d. A curve drawn through E', c, d, F' will give the plan on the edge EF.

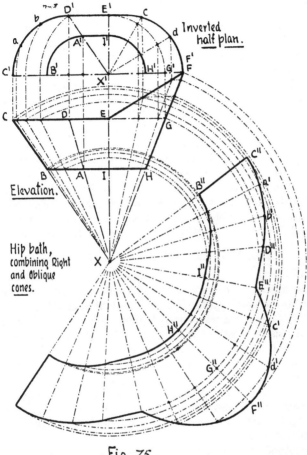

Fig. 75.

To develop the pattern on the oblique cone side, divide the quadrant $C'D'$ into three equal parts, as at $C'abD'$, and, with the apex X' as centre, swing arcs to $C'X'$. From these points drop vertical lines to CD, and from the points on CD draw lines to the apex X. Using X as centre, swing arcs into the pattern from the points on CD, and also from the corresponding points on AB, as shown in Fig. 75. Now, on the right cone side, from the points on EF, project horizontal lines to FG, the outside of the cone. Again, using X as centre, swing arcs into the pattern from the points on FG, and also from point H. It now remains to plot the pattern on the series of arcs. First, take the distances $C'a$, ab, bD', from the plan, and, beginning at a convenient spot on the arc from C, as at C'', mark them off in the pattern, stepping over from one line to the next, as shown in the diagram. Next, take the distance $D'E'$ from the plan and mark that off in the pattern, as from D'' to E''. The next step is to take the three equal divisions from the quadrant $E'G'$ in the plan, and mark them off along the inner arc in the pattern, as from E'' to G''. Now join all these points, C'',a',b',D'',E'', . . . G'', to the apex X, and in the case of those from E'' to G'' produce them outwards to meet the corresponding arcs from the side FG of the cone. A line now drawn through these points, from C'' to F'', will give the outside or top curve in the pattern. Similarly, a line drawn through the corresponding points from B'' to H'' on the inner set of arcs, as shown in the diagram, will give the bottom curve. The development up to this stage represents one half of the complete pattern. The other half is a reverse repetition on the other side of the line XF''.

TRUE LENGTHS OF THE OBLIQUE CONE

One of the most important stages in the progress of pattern developing is that which deals with the oblique cone cut by a plane not parallel to the base. In the case of the right cone all the points on the cutting plane are transferred horizontally, or at right angles to the central axis, to the outside slant of the cone, but this cannot be done with the oblique cone. Bearing in mind the fact that true lengths only are required in the pattern, it is a simple matter to obtain these on the right cone, since any point on the cutting plane taken horizontally to the outside slant, on either side, thereby presents its true distance from the apex. On the oblique cone this would, on the one side, give a length too short and on the other side a length too long.

The method of obtaining true lengths on the oblique cone is shown at (a), Fig. 76. A half-plan of the cone with the apex at A is attached direct to the base of the elevation. The line AB in the plan lies on the surface of the cone from the middle point of the semicircle to the apex. This line in the elevation is at $A'B'$, and for convenience will be called the ELEVATION LINE of AB. To obtain the TRUE LENGTH LINE of AB, the PLAN LINE is swung round from the apex A to the base line, which thus places it at right angles to the vertical height of $A'A$. The diagonal $A'B''$ is therefore the true length line. In the elevation, MN represents a cutting plane through the cone which makes an acute angle with the base. The elevation line $A'B'$ crosses the cutting plane at C'. Now, the point C', when located on the true

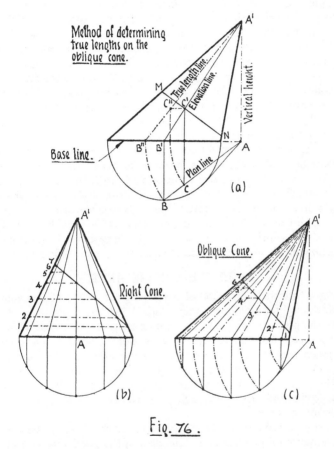

Method of determining true lengths on the oblique cone.

Base line.

(a)

Right Cone.

(b)

Oblique Cone.

(c)

Fig. 76.

length line, should be exactly the same vertical height from the base. Therefore, a horizontal line drawn from C' on the elevation line to C'' on the true length line will give the true distance of that point from the apex A'.

Alternatively, if the plan point C were swung round to the base line, and then taken vertically upwards to the true length line, the same point, C'', would be located. This method sometimes forms a convenient alternative to that of transferring C' horizontally to C'', but for most cases the latter method has many advantages, and will be used in this course as occasion arises.

The two diagrams at (b) and (c), Fig. 76, show a right cone and an oblique cone treated for true lengths from the apex to a cutting plane. In principle these two methods, although apparently different, are precisely the same. For instance, to obtain the elevation lines in the oblique cone, the division points on the semicircle are drawn vertically upwards to the base line, and then from the base line to the apex. To obtain the elevation lines on the right cone the process is exactly the same. Again, to obtain the true length lines on the oblique cone, the division points on the semicircle are swung round to the base line, using the plan apex A as centre, and then from the points on the base line the true length lines are drawn to the apex A'. In the case of the right cone, it will be observed that if the plan apex A be used as the centre for swinging round the division points on the semicircle, all of them will coincide with the outside point of the base. This means that the outside slant of the cone is the true length line for all the elevation lines. Hence the reason for transferring all the points on the cutting plane to the outside slant of the cone. If these principles be well digested, there should be no difficulty in following the developments of right or oblique conic frustums.

HOOD FOR BOILING KETTLE

Fig. 77 shows the method of developing the pattern for an oblique conical hood for a boiling kettle. These hoods are usually connected, from the pipe at the top, to a steam extraction system, which draws off the steam from the kettles. The slant of the hood depends largely on given conditions, but for ordinary purposes the example at (c), Fig. 77, with the vertical back, forms a good standard type. The method of development for this example is precisely the same as that given at (a), although the pattern would be a little different in shape.

In the example at (a) a half-plan is attached direct to the base for convenience of development. The plan of the apex falls at A on the base line inside the semicircle. To develop the pattern, divide the semicircle into six equal parts and number the points from 1 to 7. Next, from these points draw lines vertically upwards to the base line, and then from the points on the base line draw lines to the apex A' in the elevation. The lines drawn thus to the apex are the elevation lines. The next step is to determine the corresponding true length lines in the elevation. To do this, place the point of the compasses at A in the plan, and swing each of the division points on the semicircle round to the base line, and from these new points on the base line draw lines to the apex A'. The lines drawn thus to the

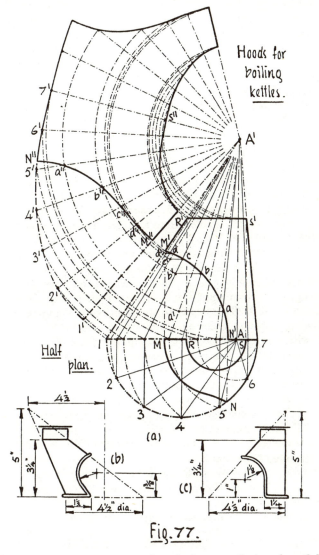

Hoods for
boiling
kettles.

Half plan.

Fig. 77.

apex from the points swung round to the base are the true length lines. Now, with A' as centre, swing out the true lengths from the base line into the pattern. Take one of the divisions from the semicircle, as 1 to 2, or 2 to 3, and, beginning at a convenient point on the outside arc from point 1, mark off the distances 1′,2′,3′,4′,5′,6′,7′, stepping over from one line to the next. A curve drawn through these points will give the base curve in the pattern for half of the full cone. From the points 1′,2′,3′,4′,5′,6′,7′, draw radial lines to the apex A'. To obtain the top curve, swing arcs into the pattern from the points where the true length lines in the elevation cross the top edge $R'S'$. Where these arcs cross the radial lines, points will occur through which to draw the top curve, as shown in the diagram at Fig. 77. This should complete a half-pattern for the full frustum.

It still remains to determine the shape of the curve $M''N''$. Crossing the curved edge $M'N'$ in the elevation are two sets of lines, the elevation lines, shown in full, and the true length lines, shown in chain dotted. To determine the true lengths for the pattern, the crossing points a,b,c,d on the elevation lines should be transferred horizontally to the corresponding true length lines. For example, taking point 5 on the semicircle in the plan, the elevation line from that point crosses $M'N'$ at a point a. A horizontal line from point a meets the corresponding true length line at point a'. Similarly, a horizontal line from the elevation line crossing at point b meets the corresponding true length at point b'. It will be observed that the true length line from point 4 in the plan coincides with the elevation line from point 3. This is due to the position of the apex A in this particular problem. Other positions would cause them to separate. Now, with the apex A' as centre swing arcs from a',b',c',d',M' into the pattern to cut the radial lines in points a'',b'',c'',d'',M''. The remaining point N'' on the base curve may be located by taking the plan distance $5N$ and marking this off along the base curve in the pattern. A curve now drawn from M'' to N'' will give the required shape of the curved edge. The full pattern is symmetrical about the line $S'',7'$. The complete form may therefore be obtained by repeating, or transferring all the points to corresponding positions on the opposite side of the line $S'''7'$.

The two examples shown below in Fig. 77, are given as additional exercises. Although the principles of development for these are exactly the same, there are numerous details of difference in form which should furnish interesting variety for practice.

AN OBLIQUE CONICAL HOPPER

The problem in Fig. 78 is a development of an oblique conical feed hopper fitting on a cylindrical screw casing. The cone in this

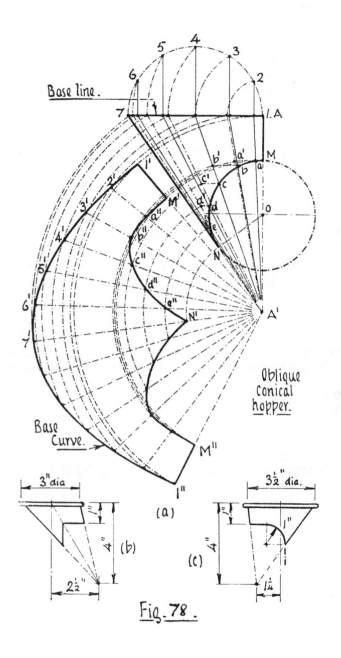

Base line.

Oblique
Conical
hopper.

Base
Curve.

3" dia

(a)

2½"

(b)

3½" dia.

4"

1¼"

(c)

Fig. 78.

case is inverted, and the apex A' falls, upwards, on the point A at one end of the base line.

To develop the pattern, describe a semicircle on the base line, and divide it into the usual six equal parts, numbering them accordingly, as at 1,2,3,4,5,6,7. For the elevation lines on the cone, drop perpendiculars from the points on the semicircle to the base line, and then from the points on the base line draw lines to the apex A'. To obtain the true length lines on the cone, with A as centre, swing arcs from the points on the semicircle to the base line, and from these points on the base line draw the true length lines to the apex A'. Now, with centre A', swing the true length lines from the base line into the pattern. Next, take one of the divisions from the semicircle, as 1,2 or 2,3, and, beginning at a convenient spot on the inside arc from point 1, mark off $1',2',3',4',5',6',7'$, to the outside arc, stepping over from one line to the next. Repeat in reverse order to $1''$. A curve drawn through these points will give the base curve in the pattern. From the points $1',2',3',4',5',6',7', \ldots 1''$, draw radial lines to the apex A'.

To determine the shape of the curved edge MN in the pattern, the crossing points of the elevation lines a,b,c,d,e should first be transferred horizontally to the corresponding true length lines, as at a',b',c',d',e', and then, with A' as centre, swing arcs from these points into the pattern, including M and N, to cut the radial lines in $M',a'',b'',c'',d'',e'',N'$. Repeat these intersections in the reverse order to M''. A curve now drawn from M' to N' and on to M'' will give the form of the curved edge MN in the pattern. When the outside edge of the cone, $7A'$, forms a tangent to the arc MN in the elevation, the exact position of the point N may be determined by drawing a line through the centre O at right angles to the tangent $7A'$. The point of intersection is the required point N.

The two examples shown at (b) and (c), Fig. 78, are given as additional exercises. It will be observed that the apex in the one case falls outside the base, and in the other case inside the base.

Chapter 6

THE PARALLEL LINE METHOD

It will be remembered that the problems of pipe intersection in the First Course were those between pipes of equal diameter. Those intersections presented straight lines in the elevation, as shown at (a), Fig. 79. An intersection between pipes of unequal diameters

Equal diameter pipes;
Straight lines of
intersection.

Unequal diameter pipes;
Curved line of
intersection.

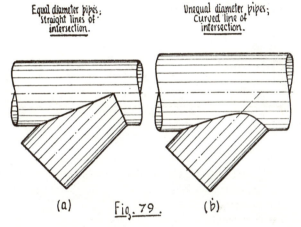

(a)   Fig. 79.   (b)

invariably presents a curved line in the elevation, as at (b), Fig. 79. In most cases this line of intersection must be determined in order to develop the pattern for the branch.

RIGHT-ANGLED TEE OF UNEQUAL DIAMETER PIPES

The example shown in Fig. 80 is of a right-angled tee of unequal diameter pipes. To develop the pattern, describe semicircles on the bases of the front and end elevations, and divide each of them into six equal parts, and number them accordingly, as in the front elevation, from 1 to 7. It will be observed that the outside point numbered 1 in the front elevation becomes the middle point in the end elevation. From these points on the semicircles draw lines perpendicular to the bases and produce them to cut the major pipe above. From the points where they cut the circle of the major pipe, from D to B, in the end elevation, project lines horizontally to meet the corresponding perpendicular lines in the front elevation. A curve drawn through the meeting points, as from *A* to *B* to *C*, will give the line of intersection.

To "unroll" the pattern, project the base line horizontally, and mark off twelve spaces, as from 1' to 1", equal to those round the semicircles. Next project the points on the major pipe circle from *D* to *B* horizontally into the pattern. From the points 1',2',3',4',5',

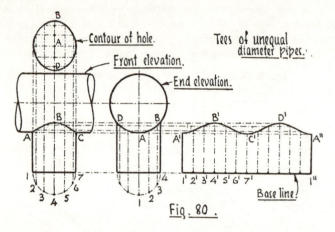

Fig. 80 .

6',7', . . . 1", on the base line in the pattern, erect perpendicular lines to meet those projected horizontally from the major pipe circle. A curve drawn through these points, *A'*,*B'*,*C'*,*D'*,*A''*, will give the contour of the intersection line in the pattern.

For the contour of the hole in the major pipe, produce the perpendicular lines in the front elevation, as shown in the diagram above the major pipe. Take the divisions round the curve from *D* to *B* in the end elevation, and mark them off along the centreline above the front elevation. Through the points thus marked, draw horizontal lines to cut the perpendicular lines from the base. A curve drawn through the points of intersection will give the contour of the hole. In this case the hole is slightly elliptical.

OBLIQUE TEE, OFF-CENTRE

The problem shown in Fig. 81 is that of a branch tee, or "stump," at an acute angle to the main pipe. This is a popular method of branching, although, more often than not, the branch is on centre with the main pipe instead of at the maximum off-centre position as shown in the diagram. However, the method of development is the same in both cases, although the patterns differ in shape.

To set out the front and end elevations, first, from any point G' on the centreline of the major pipe, set off the centre line $G'c$ of the minor pipe to the required angle. On either side of these centrelines draw lines to the required diameters of the pipes. Mark off the end

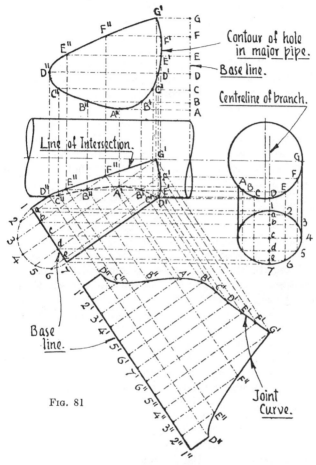

Fig. 81

1,7 of the branch pipe at right angles to its centreline. Describe a semicircle on the end of the branch piece, and divide it into six equal parts. Number the points from 1 to 7, and project perpendicular lines from these points to the end, or base, of the branch piece. Letter these points a,b,c,d,e. In the end elevation draw the circle of the major pipe, and locate the centreline of the minor pipe. Next, to obtain the elliptical end of the branch in the end elevation, project horizontal lines from the points 1,a,b,c,d,e,7, to cut the centreline of the branch in the end elevation. Now, from the semicircle in the front elevation, take each of the perpendicular distances a,2; b,3; c,4; d,5; e,6; and mark them off on the corresponding lines on either side of the centreline 1,7, in the end elevation. Through the points thus marked, draw in the ellipse. Next, draw vertical lines from the points on the ellipse to cut the circle of the major pipe above, as at A,B,C,D,E,F,G. Now, from these points on the major pipe circle, draw horizontal lines back to the front elevation to meet another set of lines drawn from the points a,b,c,d,e, parallel to the central axis cG' of the branch piece. The points of intersection, through which the line of intersection is drawn, are shown in the front elevation at $A',B',C',D',E',F',G',F'',E'',D'',C'',B'',A'$. It will be observed that points B' and B'' are on the same horizontal line projected from point B in the end elevation. Similar conditions hold good for points C',C''; D',D''; E',E''; and F',F''.

To "unroll" the pattern, extend the base line from the end of the branch, and mark off twelve divisions, as shown at 1',2',3',4',5',6',7', . . . 1″, equal to those round the semicircle. From these points draw lines at right angles to the base line. Next, from the points round the line of intersection, draw lines parallel to the base line to meet those at right angles. Now, assuming that the seam is to be on the shortest side from 1 to D'', plot the joint curve in the pattern, beginning at D'', and following each point successively round the intersection line, as shown in the illustration. It will be seen that each point on the joint curve in the pattern is assigned the same letter as the corresponding point on the line of intersection. The complete pattern may now be outlined as in the diagram in Fig. 81. The contour of the hole may be "unrolled" at right angles to the centreline of the major pipe. A base line should be erected at any convenient point, as, in this case, from the end of the portion shown. On the base line, mark off distances equal to those on the major circle in the end elevation. Thus, each of the distances AB, BC, CD, DE, EF, FG are spaced along the base line above the front elevation. From these points draw horizontal lines. Next, from the points round the line of intersection, erect vertical lines to meet the corresponding horizontals from the base line. Thus, the horizontal line from point B intersects the vertical lines from points B' and B''. Similar conditions hold for the other lines from C,D,E and F. The intersecting points obtained in this way should be sufficient to enable the contour of the hole to be drawn in.

RIGHT-ANGLED TEE, OFF-CENTRE

The problem in Fig. 82 is similar to the previous one, except that in the end view the minor pipe has a maximum off-centre position. Nevertheless, the method of development is precisely the same, and if the directions for the problem shown in Fig. 80 are carefully followed and applied to this one, there should be no difficulty in

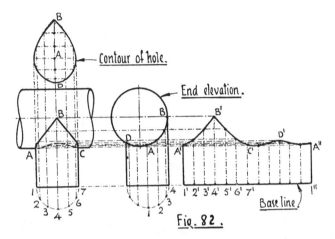

Fig. 82.

setting out the pattern for the off-centre pipe and the contour of the hole in the major pipe.

AUXILIARY PROJECTION

It often happens that drawings or blue prints, issued for the making of a system of pipe work, contain branch pieces of the type shown in Fig. 84. In most working drawings, only the plan and elevation are given, although, in some cases, end or side elevations are also included. Nevertheless, in cases like that of the branch piece shown in Fig. 84, to obtain the true line of intersection between the main pipe and the branch, yet another view is needed. In the example already given in Fig. 81, it will be seen that the line of intersection is obtained in conjunction with the end elevation. The end elevation is a view looking in the direction of the central axis of the main pipe, which thus becomes a plain circle. Lines on the branch piece intersecting this circle afford points which can easily be transferred to the corresponding lines in the front elevation. In the case of a plan and elevation as shown in Fig. 84, no such simple process is possible. In order to obtain the line of intersection a *projected* view in the direction of the central axis of the main pipe is required.

The ordinary method of representing a plan and elevation is termed orthographic projection. The plan is a view of the object seen from above, looking vertically down on it, and the front elevation is a view looking horizontally at the front. It is sometimes

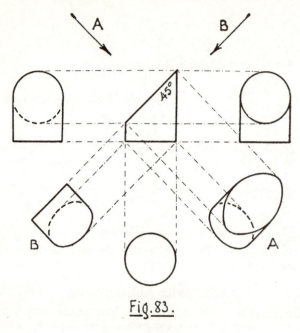

Fig. 83.

necessary to project other views in order to obtain certain essential conditions for development. In Fig. 83, the central figure is an elevation of a cylinder cut off at 45 degrees. The circle below it is a plan of the cylinder, and the diagrams on either side of the central figure are side elevations, each as viewed from the opposite side to that on which it appears. The two diagonal views at *A* and *B* are auxiliary projections which represent the cylinder as it would be seen when viewed in the directions of the arrows. There is nothing special about an auxiliary projection other than the fact it is projected at an angle which is neither horizontal nor vertical. An auxiliary projection may be taken at any angle to suit requirements.

The principles governing the setting out of the projected view are precisely the same as for the ordinary side elevation, since the

projectors are all parallel and in the direction of the angle of projection. This may be seen perhaps more clearly if the diagram shown
in Fig. 83 be rotated about the central figure so that the view
projected at *A* occupies the horizontal position. It may then be

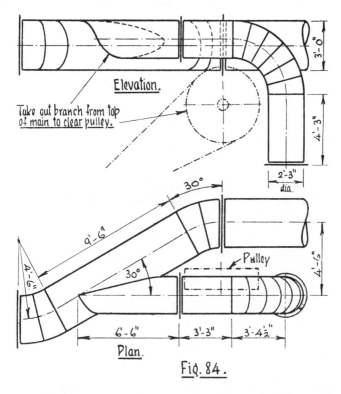

Fig. 84.

regarded as an ordinary side elevation, and the other projected
view at *B*, which is at right angles to that at *A*, becomes the plan.

In order to determine the line of intersection between the main
pipe and the branch piece shown in Fig. 84, the auxiliary projection
must first be obtained in the direction of the central axis of the main
pipe. This is shown in Fig. 85, wherein the elevation and plan of the
joint members only are given, together with the projection. The
object of projecting in the direction of the central axis of the main
pipe is to obtain that member as a plain circle, when the lines of the
branch piece intersecting that circle may be readily transferred back
to the plan view to obtain the joint line. It will be observed that this

problem, when the projected view is obtained, is precisely the same as that given in Fig. 81. The patterns may therefore be developed by following the directions given in that example.

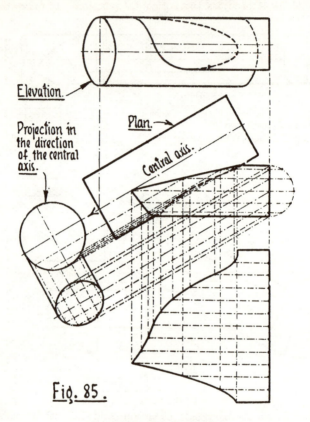

Elevation.

Projection in the direction of the central axis.

Plan.→

Central axis.

Fig. 85.

DOUBLE PROJECTION

In every problem involving pipe intersection it is necessary, in order to develop the pattern, to obtain a view wherein both of the intersecting pipes lie in a horizontal plane. That is, the axis of each pipe must lie in the plane of the paper, or the sheet. Thus, in the plan view of Fig. 85, both the main pipe and the branch lie flat on the paper, so to speak, whereas in the elevation the main pipe is inclined to the plane of the paper at an angle of 30 degrees.

Many examples in practice, as usually given in the ordinary plan and elevation of a working drawing, prove to be awkward problems of development. Very often the intersection line is shown only as an approximation, and it is left to the craftsman to work out the development accurately to meet the requirements laid down in the drawing. Although the problem itself, as in this case, may be an easy

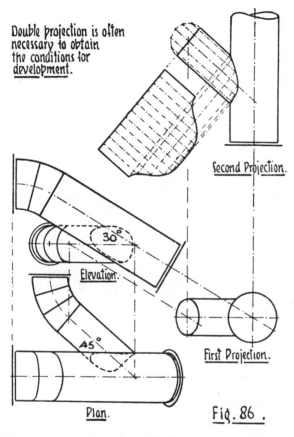

Fig. 86.

type of intersection when placed in its most convenient position, the mere shifting of one of the axes through a small angle may turn it into a complex geometrical puzzle. Nevertheless, provided that certain methods of solution are applied with a clear understanding, a good deal of time will be saved which might easily be, and often is, wasted by the application of unwise rule-of-thumb tactics.

The example shown in Fig. 86 represents, in the elevation, a main pipe dipping at an angle of 30 degrees, while the axis of the branch pipe is horizontal. In the plan, however, the branch pipe is shown entering the main at an angle of 45 degrees. It will be seen, therefore, that neither the plan nor the elevation give the required conditions for development. In order to obtain a view in which both pipes lie in the plane of the paper, a double projection will be necessary. The first projection should be in the direction of the central axis of the main pipe, which will present that pipe as a plain circle. The second projection should be at right angles to the centreline of the branch pipe in the first projection, as shown in Fig. 86. Thus, the second projection will be the view required, and the true line of intersection may be determined in conjunction with the first projection. The development of the pattern may then be obtained as in the previous examples.

This principle of double projection is one which has important application in the more advanced problems, and will be discussed in closer detail at a later stage in the course. At the moment, however, it serves to show how, in many cases, a simple problem may become difficult by the mere presentation of the conditions in the plan and elevation.

OBLIQUE CYLINDRICAL HOPPERS

The examples shown in Fig. 87 are sometimes called shoe-shaped funnels. They may also form hoppers to serve special cases. It will be seen that in each of the figures at (a), (b), (c), and (d) the base is a circle, while the top is composed of two semicircles of the same diameter with a straight side between them. The bodies are thus composed of two half-cylinders, of which one or both must be oblique with a flat triangle between them. The example at (a) is made up of one half of a right cylinder at A',B',C',D', and one half of an oblique cylinder at D',E',F',G', with flat triangles, as at C',D',E', between them.

To develop the pattern, divide the circle in the plan into six equal parts, and project the points vertically upwards to the base G',D',A'. From the points on the half-base $G'D'$ draw lines parallel to the side $G'F'$ to cut the top between C' and B'. Assuming now that the seam is to be made along A',B', begin by "unrolling" the portion of the right cylinder A',B',C',D', as shown at A'',B'',C'',D'', making the divisions from A'' to D'' equal to those in the plan circle from A to D. Next add the triangle C'',D'',E'', making it equal to the triangle C',D',E' in the elevation, but reversed in position. Now, from the points on the half-base G',D', draw lines to intersect D',E' at right angles to it. From the points obtained on D',E', draw lines horizontally to cut D'',E'' in the pattern. From the points where these lines cut D'',E'', project lines at right angles to D'',E'', as shown in the figure at (a). Next take one of the divisions between A'' and D'', and continue the spacing on the lines projected from D'',E'', by stepping over from one line to the next, as from D'' to G''.

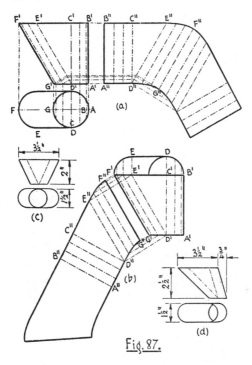

Fig. 87.

From the points from D'' to G'' draw lines parallel to D'',E'', and cut them off equal in length to D'',E''. Curves now drawn from D'' to G'' and E'' to F'' will complete half the pattern. The other half is a duplicate in reverse order on the other side of the line G'',F''. The pattern shown at (b) is for the same model, but in this case the seam is on the opposite side, from G' to F'. The smaller examples at (c) and (d) are typical variations of the same problem, and should serve as useful examples for practice.

DOWNSPOUT HEADS

Downspout heads form excellent examples of moulding sections, and usually offer a fair degree of scope for individual design Fig. 88

shows the plan and elevation of a simple type of head in five segments, three of which are equal parts of an octagonal construction.

To develop the pattern, one segment must first be arranged in plan so that the centreline is horizontal, as shown at *CL*. The true

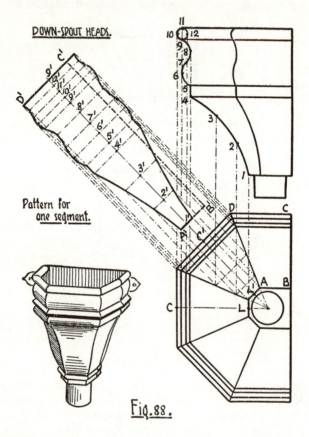

DOWN-SPOUT HEADS.

Pattern for
one segment.

Fig. 88.

shape of this segment is then given by the form of the outside line in the elevation. This line should be divided into any number of convenient parts, as in this case from 1 to 12 in the diagram. The actual length of this line is the full length of the centreline *CL*. The pattern may be projected from any of the segments in the plan, as in this case from the one above the central segment. The length of the centreline *C'L'* is the same as that of *CL*. From each of the points on the outside line in the elevation drop vertical lines to the joint line between the central segment and the one above. Project the points obtained on this joint line across to similar positions on the other side of the segment $C'L'$. For the development of this segment project the centreline $C'L'$ into the pattern, and, beginning at any convenient point, mark off the distances $1', 2', 3' \ldots 11'$, $12', 9'$, equal to those on the outside line in the elevation. Through these points in the pattern draw lines at right angles to the projected centreline. Now, from each of the points on the two joint lines on either side of $C'L'$ in the plan, project lines into the pattern parallel to the centreline. Where these cut the corresponding cross lines at right angles, points will be obtained through which to draw the contours of the two opposite edges of the pattern. Cross lines at the top and bottom between these edges will then complete the pattern.

The pattern for the end segment, A,B,C,D, in the plan may be obtained by marking off the distance AB in the pattern as shown at $A'B'$, and then drawing the straight line $B'C'$ parallel to the centre line. Then the pattern for the end segment will be contained between the contour $A'D'$ and the straight line $B'C'$.

The perspective sketch represents another typical design for a downspout head.

ROOF FINIALS

Roof finials also form a good series of examples of moulding sections and offer a wide scope for design.

The example shown in Fig. 89 is one of octagonal section, of which a half-plan is shown below the elevation. The line CL represents the centreline of the half-segment CAL. To develop the pattern, divide the contour of the outside line into any number of convenient parts, as in this case from 1 to 28, and space those distances along the centreline in the pattern. Through these points draw cross lines at right angles to the centreline. Next drop vertical lines from the points on the outside contour line in the elevation to cross the line CL and meet the joint line CA in the plan. Then take the distances between the lines CL and CA, and mark these off on both sides of the centreline in the pattern on the corresponding cross lines. Points should thus be obtained on the cross lines through which to draw in the contours of the opposite edges of the pattern.

PROBLEMS OF THE COMMON CENTRAL SPHERE

Problems of the common central sphere belong to a special class of intersections in which an imaginary sphere is the central form around which two, three, or more cones or cylinders meet or intersect

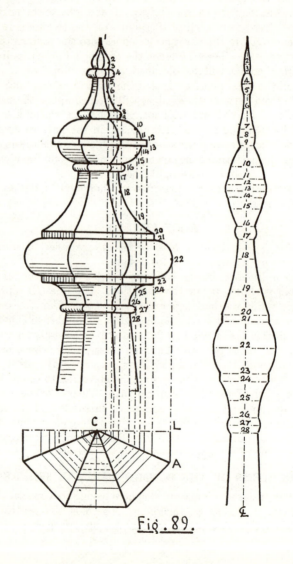

Fig. 89.

The essential condition is that the intersecting bodies should be so arranged that their surfaces, if produced, would encircle the sphere. The illustration at (d), Fig. 90, represents a revolving cowl for a chimney or fume pipe. It is a combination of a right cone and

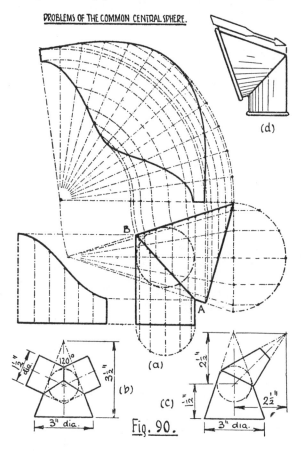

PROBLEMS OF THE COMMON CENTRAL SPHERE.

(d)

B

A

(a)

(b)

(c)

120°

1½" dia.

3½"

2½"

1½"

2½"

3" dia.

3" dia.

Fig. 90.

a cylinder. The developments are shown in the central figure at (a), and it will be seen that the sides of the cylinder just touch, or are tangential to, the central sphere. The sides of the cone also just touch the central sphere. The important point by these conditions is that the line of intersection in the elevation becomes a straight line, as from A to B. The smaller illustration at (b) shows a right cone intersected by two cylinders, and it will be observed that the sides of the cone and also the sides of the cylinders form tangents to the central sphere. The intersecting lines are therefore represented as straight lines. Similar conditions hold good for the example at (c), which represents a right cone intersected by another right cone in such a way that the sides of both cones form tangents to the common central sphere. The intersection line thus becomes straight.

The methods of developing the patterns for the cowl in the central figure at (a) have already been dealt with. The conical portion is similar to the examples given in Figs. 7 and 9, and the cylindrical part is the same as that shown in Fig. 22.

The smaller figures at (b) and (c) are given as additional exercises.

CHAPTER 7

TRIANGULATION

ADAPTABILITY OF TRIANGULATION

MOST of the problems of pattern-drafting fall to the method of triangulation for solution. Of the three methods by which all patterns may be developed triangulation is by far the most adaptable. The Radial Line method and the Parallel Line method can only be used on their own particular class of problem, but the method of triangulation is serviceable, not only on those problems which cannot be solved by other means, but also on those which rightly belong to the other two classes. For instance, many examples of work which are formed of cones, either right or oblique, may be more conveniently developed by the method of triangulation if the distance from the apex to the base involves radii of unusual length. Nevertheless, this convenience should not detract from the use of the Radial Line or the Parallel Line method, since any pattern drafted by either of these methods is likely to be more accurate than the alternative development by triangulation.

The method of triangulation is a process of building up the pattern piece by piece in the form of triangles. The principles were fully explained in the First Course, but it may be well to repeat here that the true size of each triangle is found by obtaining the true length of each side, and marking it off in its relative position in the pattern. The true length of any line may be found by taking its length from the plan—called its *plan length*—and placing it at right angles to its vertical height obtained from the elevation. The diagonal between the free ends will then give the true length of the line. It is important to observe that, although the plan length is the actual length taken from the plan, the vertical height is not necessarily the actual length represented in the elevation. For example, in Fig. 92, the line 3',6' in the elevation is leaning at an angle to the horizontal. Its vertical height is not the distance between 3' and 6', but the vertical depth from the top point 6' to the horizontal plane of its base, as from 6' to B on the vertical height line.

The problems of triangulation dealt with in the First Course were of such objects as lay between two parallel or horizontal planes. This

113

condition premised that all the vertical heights in any one problem should be the same, which somewhat simplified the work, since only one vertical height was needed. In this Second Course, the objects dealt with will lie between planes inclined at an angle to each other,

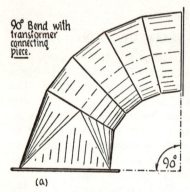

90° Bend with transformer connecting piece.

(a)

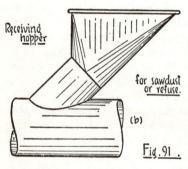

Receiving hopper

for sawdust or refuse.

(b)

Fig. 91 .

which will involve at least two, and sometimes many, different vertical heights. A little more care will be needed in following up the solution, but once the idea is fully grasped and the principles understood, there should be no difficulty in solving any problem of this kind.

The illustration at (a), Fig. 91, shows a square-to-circle transformer occupying the position normally taken by the other two segments of the full right-angled bend. A combination of this kind is often required on the square or rectangular outlet of a fan. The illustration at (b), Fig. 91, shows a receiving hopper for sawdust or refuse which is delivered to the duct below and is drawn from there by a fan to be collected in a cyclone separator. These are representative of only a few applications of this type of transformer.

RECTANGLE-TO-CIRCLE TRANSFORMER WITH CIRCLE INCLINED

The example shown in Fig. 92 has a rectangular base and a circular top inclined at an angle to the base. The circular hole at the top becomes an ellipse in the plan. To obtain this ellipse, describe a semicircle on the top line 2′,10′, in the elevation, and divide it into six equal parts. Number the points 2′,4′,5′,6′,8′,9′,10′, and project them perpendicularly back to the line 2′,10′. Number these points also 2′,4′,5′,6′,8′,9′,10′. From the latter points on the line 2′,10′ drop verticals into the plan to cut through the centreline 2,10. On either side of the centreline 2,10 in the plan, mark off distances above and below equal to the corresponding distances from 2′,10′ to the semicircle in the elevation. Sufficient points should thus be

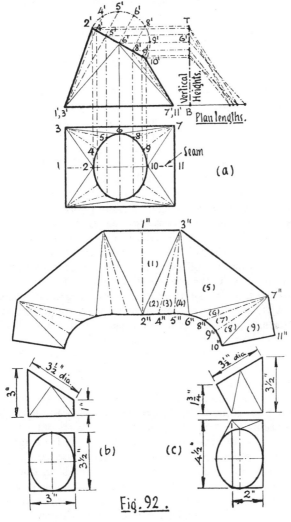

Fig. 92.

afforded through which to draw in the ellipse. Assuming now that the seam is to be on the middle of the short side, as from 10 to 11, number the points in the plan in accordance with the method explained in the First Course, as shown here at 1,2,3,4, . . . 10,11.

Only half of the plan is numbered, as the other half is similar and symmetrical about the horizontal centreline 1,11. Next, erect a vertical height line BT from the base produced, and project all the points on the top edge 2',10', horizontally to the vertical height line.

Now, to develop the pattern, take the plan length from 1 to 2, and mark this distance off along the base line from the bottom point of the vertical height line B. The diagonal from the point marked off on the base line to the point on the vertical height line corresponding to point 2' from the elevation will give the true length. Take this true length and mark it off in any convenient position, as at 1",2", in the pattern. Next take the plan length from 2 to 3 and mark this off along the base line at right angles to the vertical height. Take the true length diagonal, and from point 2" in the pattern swing off an arc through point 3". Next take the distance 1,3 direct from the plan and from the point 1" in the pattern describe an arc cutting the previous arc in 3". Lines joining these three points should complete the first triangle. It is important to note that the distance 1,3 in the plan is a horizontal line and has no vertical height. It is therefore already a true length and does not need triangulating against the vertical height line. For the next triangle take the plan length 3,4, and mark it off along the base line at right angles to the vertical height. Take the true length diagonal, this time being careful to observe that the point on the vertical height line corresponds to point 4' in the elevation. From point 3" in the pattern swing off an arc through point 4". Now, since the next distance 2",4" in the pattern must be the true distance between those points, that length is not obtained from the ellipse in the plan, because those spacings are somewhat foreshortened. The required true distances for the spacings around the circular top should be taken from the semicircle on 2',10' in the elevation. Thus, take the distance 2',4' from the semicircle and from point 2" in the pattern describe an arc cutting the previous arc in 4". Join 2",4' and 3",4". This should complete the second triangle. For the third triangle take the plan length 3,5 and triangulate it against the vertical height, being careful that the vertical height taken corresponds to the point 5' in the elevation. Take the true length diagonal and from point 3" in the pattern swing off an arc to cut the point 5". From the semicircle in the elevation take the distance 4',5', and from point 4" in the pattern describe an arc cutting the previous arc in point 5". For the fourth triangle repeat this process with plan length 3,6 and the distance 5',6' from the semicircle. For the fifth triangle, repeat with the plan length 6,7, and also the true plan length 3,7, but note in this case that the triangle is reversed in position. The remainder of the pattern should be straightforward and easy to follow from this point, since the process is the same to the end. The other half of the pattern on the other side of the centreline 1″,2″ is a repetition in the reverse order.

The smaller examples at (b) and (c) are given as additional exercises.

RECTANGLE-TO-CIRCLE TRANSFORMER WITH RECTANGLE INCLINED

In the previous example the circular top was inclined at an angle to the horizontal base. In this problem, Fig. 93, the top is horizontal, while the base is inclined at an angle of about 40 degrees. These two examples may be regarded as typical of a class in which there are many minor variations, but the method of development is precisely the same.

The chief difference in these two problems lies in the vertical heights. It will be observed that in the previous example seven different vertical heights are encountered as the points taken occur down the slope of the top edge, but in this problem only two vertical heights are needed which correspond to the positions of the ends of the inclined base.

To develop the pattern, divide the circle in the plan into the usual twelve equal parts, and, assuming that the seam is to be on the middle of the short side, as from 10 to 11, number the points in the order shown in the figure. Erect a vertical height line BT, and project horizontal lines from the two ends of the base to serve as base lines for the plan lengths.

For the first triangle, take the plan length 1,2, and mark it off along the bottom base line from the point B. Take the true length diagonal to the top point T, and mark it off in any convenient position, as at 1′,2′ in the pattern. Next take the plan length, 2,3 and triangulate it against the vertical height. Take the true length diagonal and from point 2′ in the pattern swing off an arc through the point 3′. Now take the true length 1,3 direct from the plan, and from point 1′ in the pattern describe an arc cutting the previous arc in point 3′. For the second triangle, take the plan length 3,4, and triangulate it against the vertical height. Take the true length diagonal and from point 3′ in the pattern swing off an arc through the point 4′. The true distance between 2 and 4 may, in this case, be obtained direct from the plan. Therefore, take the true length 2,4 from the plan, and from point 2′ in the pattern describe an arc cutting the previous arc in point 4′. The remainder of the pattern should be easily followed from this point, since the process of development is the same throughout, and the directions merely a repetition

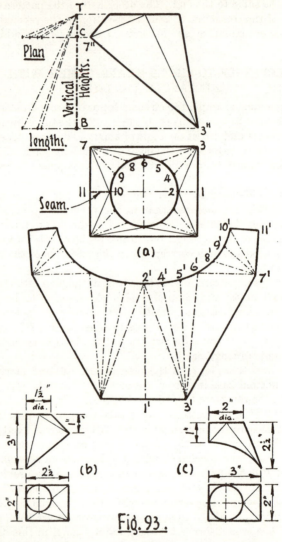

Fig. 93.

of terms. Nevertheless, a word of caution may be needed when the fifth triangle is arrived at; the true distance 3',7' in the pattern is not obtained in this case from the plan, but from the elevation. Thus, the distance from 3″ to 7″ is the true length required. It is also important to observe that those plan lengths which terminate on the higher side of the rectangle at points 7 and 11 should be marked off along the corresponding top base line at right angles to the short vertical height C,T.

The two smaller figures at (b) and (c) are given as additional examples for practice. In the case of that shown at (b) it will be seen that the circular top is off-centre both ways, and is not symmetrical about any axis. Therefore, the whole of the pattern has to be developed. The example at (c) is symmetrical about the horizontal centreline in the plan.

PROJECTED VIEWS

The fact that working drawings often contain problems of development presented in the most awkward manner is a circumstance

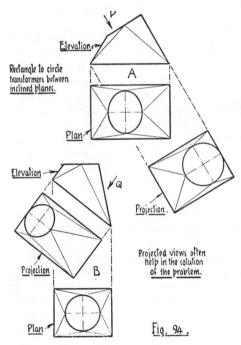

with which many sheet metal workers are familiar. It has already been shown in a previous chapter that the plan and elevation of an object, seen as it would be in position at erection, do not always form the best views for developing the pattern. The principles of auxiliary projection may often help the craftsman to simplify the work of development, and the skill of the pattern-drafter is revealed in the method he adopts in tackling these problems.

Although the tallboy transformers dealt with in the two previous examples may be regarded as two separate types, each may be transformed into the other by means of an auxiliary projection. In

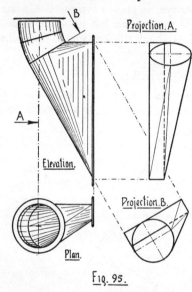

Fig. 94, the plan and elevation at A represent the type in which the base is horizontal and the circle is inclined; but if a projected view in the direction of the arrow P be taken, and the views turned so that this may be used as the plan, then the problem becomes one in which the circle is horizontal and the base inclined. Alternatively, the plan and elevation shown at B represent the type in which the circular top is horizontal and the base inclined, but if a projected view in the direction of the arrow Q be taken, and the views turned so that this may be used as the plan, then the problem becomes one in which the base is horizontal and the circle inclined.

The reversibility of these two problems by auxiliary projection is an important principle which should be well studied and fully grasped.

An auxiliary projection will sometimes simplify the work of developing the pattern. An example of this will be seen in Fig. 95, which represents a transformer connection from a vertical rectangular hole to the segment of a lobster-back bend. The plan and elevation are such as would be given in a working drawing, but if an attempt were made to develop the pattern from these two views, the process would be troublesome owing to the awkwardness of the vertical heights. Two alternative projections are shown, one in the direction of the arrow A, and the other in the direction of the arrow B. The projection A, if used as a plan, presents the type with a horizontal base and circle inclined. The projection B, if used as a plan, presents the alternative problem with circle horizontal and base inclined. This latter projection at B is the better for the purpose of development, and the pattern is shown developed in Fig. 96.

TRANSFORMER CONNECTION TO LONG RECTANGULAR HOLE

To develop the pattern for the transformer connection shown in Fig. 96, divide the circle in the plan into twelve equal parts, and join the points in each quadrant to the corresponding corner point on the rectangle. Assuming that the seam is to be on the middle of the short side, as from 1 to 2, number the points in accordance with the prescribed method, as shown in the plan. Erect a vertical height line BT, and project the base lines from each end of the rectangle, as at B and C. For the first triangle, take the plan length 1,2, and mark this distance off along the top base line at right angles to the short vertical height CT. Next take the true length diagonal to the top point T, and mark it off in any convenient position, as at 1',2', in the pattern. Next take the plan length 2,3, and mark it off along the baseline from C. Take the true length diagonal, and from point 2' in the pattern swing off an arc through the point 3'. Now take the true length 1,3 direct from the plan, and from point 1' in the pattern describe an arc cutting the previous arc in point 3'. Join 1',3' and 2',3'.

For the second triangle, take the plan length 3,4, and triangulate it against the vertical height. Take the true length diagonal and from point 3' in the pattern swing off an arc through point 4'. Take the true distance between 2 and 4 direct from the plan, and from point 2' in the pattern describe an arc cutting the previous arc in point 4'. Join 2',4' and 3',4'. This should complete the second triangle. For the third triangle, take the plan length 3,5, and triangulate it against the vertical height. Take the true length diagonal and from point 3' in the pattern swing off an arc through point 5'. Next take the true distance 4,5 direct from the plan, and from point 4' in the pattern describe an arc cutting the previous arc in point 5'. For the fourth triangle, repeat this process with plan lengths 3,6 and 5,6. For the fifth triangle, again repeat with plan lengths 3,7 and 6,7, but note in this case that the triangle is reversed in position, also that the true length 3',7' should be obtained direct from the elevation, as from 3″ to 7″. The remainder of the pattern should be easy to follow from this point, since the process of development is the same throughout. Nevertheless, it may be well here to

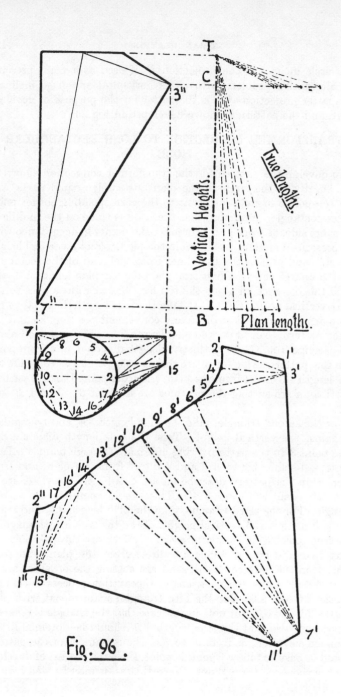

True lengths.

Vertical Heights.

Plan lengths.

Fig. 96.

note that the plan lengths from the corner points 7 and 11 should be triangulated against the full vertical height *BT*.

TRANSFORMER SEGMENT

The example shown in Fig. 97 represents one method of connecting a square and a circular hole at right angles to each other by means of the transformer segment in the middle. The square pipe, cut off at an angle, forms a rectangle at the base of the segment, and the circular pipe, cut off at an angle, forms an ellipse at the top of the segment. The size and shape of this ellipse may be obtained by describing a semicircle on the end of the circular pipe, and dividing it into six equal parts. Next project these back at right angles to the end of the circular pipe, and on to the top of the segment 2′,10′. From the points on the top edge of the segment, draw lines at right angles to it, and cut them off equal in length to the corresponding distances from the end of the pipe to the semicircle, thus affording the points 2′,4′,5′,6′,8′,9′,10′, through which to draw in half of the ellipse.

For the purpose of developing the pattern for the middle segment a projection will be advisable, either at right angles to the rectangular base or at right angles to the elliptical top. In the example given, the projection is at right angles to the base, and is used as a plan. The vertical height line *BT* is erected also at right angles to the base, and the plan lengths marked off along the base line, as shown in the figure. The problem is now one of the type in which the base is horizontal and the top inclined, the top in this case being an ellipse. The directions for developing the pattern are precisely the same as for the problem shown in Fig. 92, except that the true distances round the top edge in this case are the spacings round the semi-ellipse instead of a semicircle. It should be noted that these spacings are not all equal, as they are in the case of the semicircle.

STOVE CHIMNEY CONNECTION

The outlets from gas and coke stoves often take the form of a rectangular stump with semicircular ends. These outlets are some-times connected to circular chimneys or flues. The connecting piece, usually of sheet metal, is a transformer from the one shape to the other.

The problem shown in Fig. 98 is typical of a chimney connection from the stove in the cabin of a canal boat. The outlet stump as a rule is on the top of the stove, and the chimney, usually a short one,

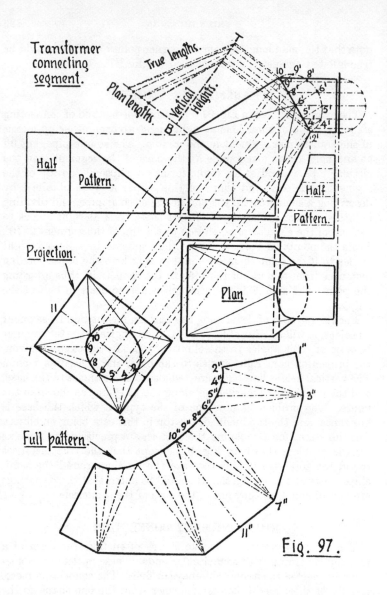

Transformer connecting segment.

True lengths.

Plan lengths.

Vertical Heights.

Half Pattern.

Half Pattern.

Projection.

Plan.

Full pattern.

Fig. 97.

passes up through the roof or out through the side of the cabin. The
latter case is represented in Fig. 98, in which the short, almost horizontal piece at the top of the transformer is the end of the cylindrical
chimney pipe inside the cabin.

The joint line 2,14 in the elevation is obtained by bisecting the

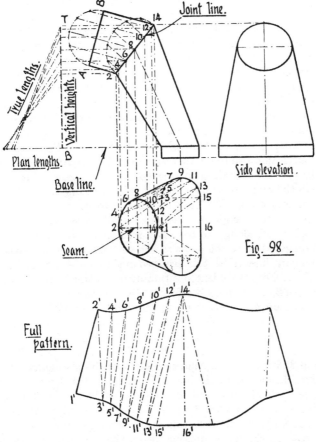

Fig. 98.

angle between the centrelines of the cylindrical pipe and the transforming piece. Since this cuts the cylindrical pipe at an angle, the
actual shape at the joint is an ellipse. To obtain the true shape of
this ellipse, describe a semicircle on the cross-section at AB, divide
it into six equal parts, and from the points on the semicircle project
lines perpendicularly back to AB and on to the joint line. From the

points thus obtained on the joint line, draw lines at right angles to it and cut them off equal in length to the corresponding distances between the line AB and the semicircle. Points should thus be afforded through which to draw in the half-ellipse as shown above the joint line 2,14. The plan, omitting the cylindrical pipe, is that of the transforming piece only. The elliptical joint line also forms an ellipse in the plan, and is obtained by dropping vertical lines from the joint line 2,14 in the elevation to cut the centreline 2,14 in the plan and marking off distances above and below the centreline equal to those between the line AB and the semicircle.

In preparing to develop the pattern, divide half of the plan into triangles as shown in the diagram, beginning at the seam 1,2 on the shorter side of the transformer, and number the points in accordance with the prescribed method. It will be observed that the consecutive numbers pass alternately from bottom to top and top to bottom, thus forming the zigzag line, which is best seen in the pattern. It will be advisable also to number the points 2,4,6,8,10,12,14 along the joint line in the elevation to correspond with those points on the ellipse in the plan. Next, erect a vertical height line BT and project all the points 2,4,6,8,10,12,14 horizontally to it.

For the first triangle in the pattern, take the plan length 1,2, and mark it off along the base line from point B. Take the true length diagonal up to the point on the vertical height line level with 2 on the joint line. With this distance in the compasses mark off the line 1',2' in any convenient position in the pattern. Next take the plan length 2,3, and mark it off from B along the base line. Take the true length diagonal, again up to the point level with 2, and from point 2' in the pattern, swing off an arc through point 3'. Now take the true length 1,3 direct from the plan, and from point 1' in the pattern describe an arc cutting the previous arc in point 3'. Join 1',3' and 2',3'. For the second triangle take the plan length 3,4 and mark it off from B along the base line. Take the true length diagonal, this time up to the point on the vertical height line level with 4 on the joint line. From point 3' in the pattern swing off an arc through point 4'. Now, to complete this second triangle, the distance 2',4' in the pattern should be obtained from the divisions on the ellipse above the joint line, since those distances are the true spacings required. Take the distance 2,4 from the ellipse, and from point 2' in the pattern, describe an arc cutting the previous arc in point 4'. For the third triangle, take the plan length 4,5, and triangulate it against the vertical height. Take the true length diagonal up to the point level with 4, and from point 4' in the pattern swing off an arc through point 5'. Next take the true distance 3,5 direct from the plan, and from point 3' in the pattern describe an arc cutting the previous arc in point 5'.

The remainder of the pattern should be easily followed from this point by carefully picking out the correct plan lengths and true distances. It may be well to observe before passing on that all those plan lengths which pass from top to bottom and bottom to top, such as 1,2; 2,3; 3,4; 4,5, and so on, should be triangulated against the vertical height. All those plan lengths around the bottom edge, such as 1,3; 3,5; 5,7; 7,9, and so on, should be used direct from the plan, as they are already true lengths. All those spacings around the top edge, such as 2,4; 4,6; 6,8; 8,10, and so on, should be obtained from the true divisions on the ellipse above the joint line. Only half of the pattern is shown developed in the figure. Since the pattern is symmetrical about the line 14',16', the other half is a repetition in the reverse order on the other side of that line.

STOVE CHIMNEY, OFF-CENTRE

The problem shown in Fig. 99 represents the alternative case of the chimney passing up through the roof of the cabin. The plan shows the position of the uptake as being off-centre both ways above the stove outlet. In order to develop this pattern satisfactorily a projected view is advisable in the direction of the arrow, which is at right angles to the line joining the centre of the base and the top. The views should then be turned so that the projection may be used as the elevation. The position of the joint line in the projected elevation is obtained by bisecting the angle between the centrelines of the transformer and the vertical pipe.

The directions for developing the pattern are the same as for the previous example, except that the whole of the pattern has to be triangulated as it is not symmetrical about any line.

STOVE ELBOW CONNECTION

The problem shown in Fig. 100 represents an elbow connection from the outlet at the back of a stove. The portion at A fits over the outlet, which is rectangular with semicircular ends. The portion at B transforms from the joint at the elbow to the circular top, which may then be connected to a pipe or chimney. A projected view of the joint line in the direction of the arrow P is shown below the side elevation. This projection is the true shape of the joint line. The plan below the front elevation is of the transformer B only. It will be seen that the shape of the joint line in the plan is similar to

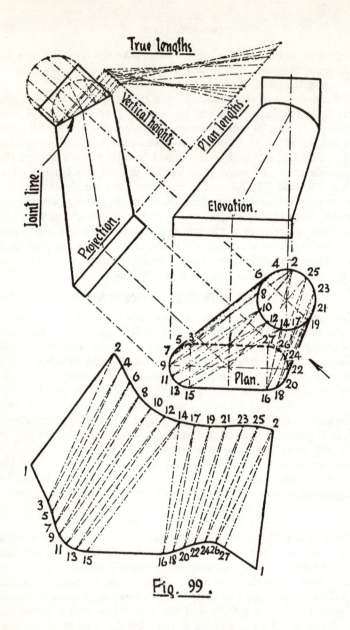

True lengths

Vertical heights.

Plan lengths.

Joint line.

Projection.

Elevation.

Plan.

Fig. 99.

that of the projection, but not so wide. It is, in fact, similar to a
projection in the direction of the arrow Q.

In preparation for developing the patterns, divide one of the
semicircular ends in the front elevation into six equal parts, as shown

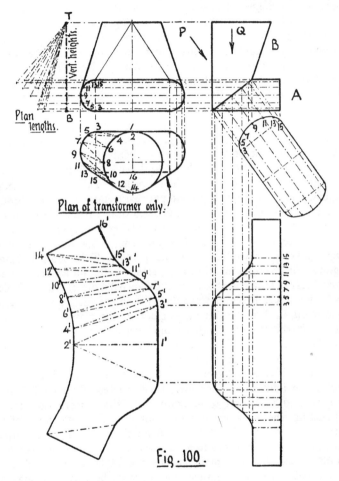

Fig. 100.

at 3,5,7,9,11,13,15. Through these points draw horizontal lines
right through to the end of the connecting piece at A. These lines
will cut the other semicircle in similar points to the first one. From
these points on both semicircles drop vertical lines into the plan to
obtain the points 3,5,7,9,11,13,15, on the base joint line. Divide

half of the top circle into six equal parts to correspond with the six parts obtained at one end of the base. It will be seen that the plan is symmetrical about the vertical centreline, so that only half of the plan need be triangulated for the pattern. Number the points as shown in the diagram, passing in consecutive order from bottom to top and top to bottom, thus obtaining the zigzag line which divides the surface into triangles. The first line 1,2 passes vertically up the centre of the back of the transformer, and is therefore represented by a single point. Erect a vertical height line BT, and project each of the points 3,5,7,9,11,13,15 in the elevation horizontally to cut the vertical height line, and produce them to form base lines along which to mark off the plan lengths.

For the first triangle, since the line 1,2 has no plan length, the full vertical height BT will give its true length. Take this height in the compasses and mark off 1′,2′ in any convenient position in the pattern. Next take the plan length 2,3, and mark it off along the bottom base line level with point 3 in the elevation. Take the true length diagonal and from point 2′ in the pattern swing off an arc through point 3′. Now take the true length 1,3 direct from the plan, and from 3′ in the pattern describe an arc cutting the previous arc in 3′. For the second triangle, take the plan length 3,4, and mark it off along the bottom base line level with point 3. Take the true length diagonal, and from point 3′ in the pattern swing off an arc through point 4′. Next take the true length distance 2,4 direct from the plan, and from point 2′ in the pattern describe an arc cutting the previous arc in point 4′. For the third triangle, take the plan length 4,5, and mark it off, this time on the base-line level with point 5 in the elevation. Take the diagonal true length and from point 4′ in the pattern swing off an arc through point 5′.

The true distance between 3 and 5 is obtained from the projected view of the joint line. Therefore, take the true distance 3,5, from the projected view and from point 3′ in the pattern describe an arc cutting the previous arc in point 5′. The remainder of the pattern should be easily followed from this point since the directions are merely a repetition of terms. The chief points to be careful of are that the plan lengths should be marked off along the correct base lines against the vertical height, and that the true lengths between 3′ and 15′ should be taken from the projected joint line.

The pattern for the horizontal portion of the elbow is shown "unrolled" below the side elevation. It is a straightforward example of Parallel Line development.

HOPPERS WITH KINKED SIDES

Some things are not so simple as they look. The hopper for feeding
the boot of the elevator shown in Fig. 101, (a) and (b), appears to be
a straightforward piece of work, yet it is the type of job which often
causes a deal of trouble unless its geometry is properly understood.
The two sides or cheeks contain a slight bend or kink from one top

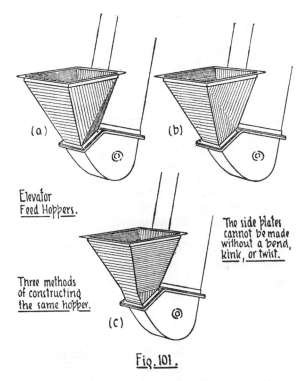

Elevator
Feed Hoppers.

Three methods
of constructing
the same hopper.

The side plates
cannot be made
without a bend,
kink, or twist.

Fig. 101.

corner to the diagonally opposite bottom corner. The kink may be
made on either one diagonal or the other, but the kink will have the
crown of the bend either inwards or outwards, "knuckle in" or
"knuckle out," according to which diagonal is chosen. The illustration at (a), Fig. 101, shows the kink from the top back corner
to the bottom front corner. In this case the kink will be "knuckle
out." The illustration at (b), Fig. 101, shows the kink from the top
front corner to the bottom back corner. In this case the kink will be
knuckle inwards.

This looks simple enough, but many draughtsmen fail to show that kink on the drawing, and many craftsmen fail to see that it must be there. Moreover, when this necessity is fully realized, trouble often arises through incorrect bending, and many a kink has had to be reversed in order to put it right. The alternative patterns for this hopper are slightly different according to the choice of kink, so that it is important to be clear on this point before developing the pattern. The observance of these little differences makes all the difference between a job which "goes right" and a job which is tantalizingly out a bit here and a bit there. Lack of foresight in this respect can easily double the cost of the job.

The example shown at (a), Fig. 101, has the advantage of giving a slightly larger capacity to the hopper than the alternative construction at (b). The example shown at (c) is yet another method of constructing the same hopper, but this construction gives a smaller capacity than that shown at (b). The kinks in the sides of the hopper at (c) are horizontal, or parallel with the top of the hopper, and the small triangles below are in vertical planes. This construction has something to be said in its favour, as some craftsmen hold the view that it gets rid of the twist, but actually this is done by putting the kink across the bottom parallel to the top. The remaining part of the side above the kink is then quite flat.

FEED HOPPER TO ELEVATOR BOOT

The pattern shown in Fig. 102 is that of a hopper similar to the example at (a), Fig. 101. The development is obtained by straightforward triangulation, although there are one or two points in the process which perhaps need explaining.

To develop the pattern, first set down the plan and elevation to the required dimensions. The kink will be required from the back top corner to the bottom front corner, as from 5 to 4. This, then, should be put in as the diagonal for that side of the hopper. Since the hopper is symmetrical about the horizontal centreline in the plan, only half of it need be triangulated. Beginning at the centreline, number the points and corners from 1 to 8 as shown in the diagram. A half side elevation is given in order to show the triangulation from 5 to 8. Project base lines from points 2 and 8 in the elevation and erect a vertical height line as shown in the figure.

Perhaps in this case the simplest way to begin the pattern is by first setting out the full front plate in the following manner. Take the true length of the centreline of the front plate, which is the distance from 1 to 2 in the front elevation, and mark off $1', 2'$ in the pattern. Draw 1',3' at right angles to it, and also 2',4', each equal to 1,3 and 2,4 respectively from the plan. Produce these on the other side of the centreline in the pattern, as from 1' to 3" and 2' to 4". Join 3',4' and 3",4". Next take the plan length 4,5 and mark it off along the bottom base line at right angles to the vertical height.

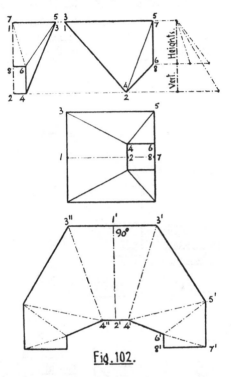

Fig. 102.

Take the true length diagonal, and from point 4' in the pattern swing off an arc through 5'. Now take the true length distance 3,5 direct from the plan, and from point 3' in the pattern describe an arc cutting the previous arc in 5'. For the next triangle take the plan length 5,6 and mark it off along the top base line at right angles to the vertical height. Take the true length diagonal and from point 5' in the pattern swing off an arc through point 6'. Next take the true length 4,6, this time direct from the front elevation, and from point 4' in the pattern describe an arc cutting the previous arc in 6'. The remaining part of the pattern may either be obtained by triangulation or by taking the true distances direct from the side elevation.

Thus, in the side elevation, portion enclosed by the line 6,5,7,8 is its true size and shape. Therefore the lines 6,7 ; 5,7 ; 7,8 ; 6,8 may all be taken direct from the side elevation and marked off in the pattern.

DELIVERY CHUTE FROM ELEVATOR HEAD

The delivery chute from the head of the elevator presents similar problems in the matter of twisted sides. The illustration at Fig. 103 shows such a chute leaving the elevator head at an angle to feed into

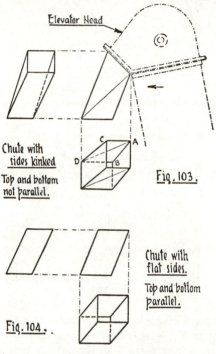

a forward position off-centre with the elevator. Thus the chute slopes both ways. The sides or cheeks of the chute will have to be kinked along one or the other of the diagonals in order to fulfil the conditions shown in the plan, elevation and side elevation of the figure. Now, it will be seen from the plan that the material ejected from the elevator head will flow chiefly down the corner of the chute marked AB. It would be a disadvantage, therefore, if this corner were sharp or acute. If the kink were made from A to D, as shown, this corner preserves its maximum angle. It is clearly an advantage, therefore, to kink the side from A to D.

Craftsmen sometimes have difficulty in deciding whether the sides of any particular chute should be kinked or not. The best guide is a careful study of the plan and elevation, bearing in mind the following points. If the top and bottom of the chute are parallel, as in the example shown in Fig. 104, then, even though the chute be off-centre both ways, all sides will be flat. If the top and bottom of the chute are not parallel, as in Fig. 103, then there is only one position which will give flat sides, and that is when the chute is on-centre with the elevator head and the sides are in vertical planes. Whenever the sides lean sideways to or from the head, kinks will be necessary.

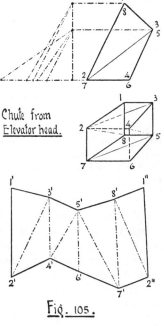

Chute from Elevator head.

Fig. 105.

The problem given in Fig. 105 shows the development of the pattern for a chute similar to that illustrated in Fig. 103. If the principles of triangulation have been carefully followed up to this point there should be no difficulty in solving this problem. It will be seen, however, that in triangulating the surface, the usual zigzag line is modified slightly in order to bring the two kinked diagonals 2,3 and 5,7 into similar positions on each side of the chute. For the rest, this problem is given as an additional exercise.

THIRD COURSE

CHAPTER 8

THE RADIAL LINE METHOD

NEARLY all complicated drawings are merely aggregates of simple parts. Each simple part may be easy to understand, but it is the fitting of the components together in their proper relation to each other which often makes the result appear somewhat complex. Similarly, an intricate pattern-development is merely an aggregation of simple processes, and these simple processes merged together in the complete solution sometimes make the problem appear difficult. However, when these simple processes, or principles, are followed in their right order, it is surprising how the difficulties melt away.

This is the beginning of a third course in pattern development, and the work is based on the following arrangement—

(a) developments by the Radio Line method of patterns for right conical connecting pieces, and multiple branch pieces for pipe work involving oblique conic frustums;

(b) developments by the Parallel Line method of patterns for branch connections on lobster-back bends, and moulding patterns for ornamental bowls and vases;

(c) further intersections of the right cone and cylinder around a common central sphere; and

(d) developments by the method of triangulation of patterns for transformers, comprising hoppers, hoods and connecting pieces, fitting on plane, curved, and angular surfaces; also multiple branch pieces and junctions for air duct work.

CONICAL VENTILATOR BASE

Occasionally the base of a ventilator is required in the form of a conical connecting piece between the cylindrical pipe above and the square section below. The illustration at Fig. 106 is a perspective view of such a connecting piece, which appears to be a straightforward conical construction, but when given in ordinary plan and elevation, it is often confused with the common square-to-circle tallboy. A glance at the half-plan, in Fig. 108, will show the similarity in that view, and the error which is sometimes made is to divide up the plan in accordance with the method of triangulation applied to the tallboy.

To set out the plan and elevation, first draw the circle in the plan equal to the diameter of the cylindrical pipe. Next draw the square concentrically round it to the size required. Since the figure is a right cone cut by four vertical planes, forming the square in the plan, the diameter of the base of the cone will be equal to the diagonal of the square. Therefore, draw the circle touching the points of the square to represent the base of the cone. Next draw the elevation of the full cone as shown at A',B,C, and insert the two vertical cutting planes corresponding to the left- and right-hand sides of the square. To obtain the front cutting plane in the form

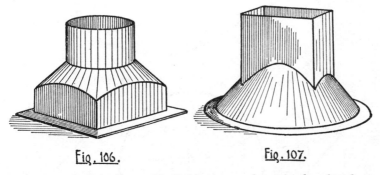

Fig. 106.   Fig. 107.

of the hyperbolic curve e,g',h, divide the quadrant in the plan from C to G into four equal parts, as shown at C,D,E,F,G, and join the points to the plan apex A. From the points D,E,F, on the quadrant, draw lines vertically upwards to meet the base B,C in points d,e,f. Join these points, also point A, to the apex A' in the elevation. The position of point g' on the centreline A,A' will be equal in height to point n on the slant of the cone. Transfer point n horizontally to meet the centreline in g'. Now the line F,A in the plan is the same as line f,A' in the elevation. The line F,A in the plan cuts the vertical cutting plane H,E in point f'. Transfer this point f' vertically upward to meet the elevation line f,A', in point f''. This gives another point on the curve e,g',h. A similar point h'' may be obtained by transferring point f'' to an equal distance on the other side of the centreline A,A'. More points on the curve may be obtained in a similar manner by dividing the arc G,F,E into a greater number of parts. A curve drawn through these points will give the required form of the cutting plane in the elevation.

To develop the pattern, first obtain the outline pattern of the full cone by swinging out the arc C',C'' from the apex A' with a radius equal to the slant of the cone. Since the quadrant C,G is divided into four equal parts instead of the usual three, mark off sixteen of these parts round the arc in the pattern. The first four, corresponding to those in the quadrant, are lettered C',D',E',F',G'. Join all these points in the pattern to the apex A'.

Now, since this problem is of a right cone, the true lengths of the lines A',g' and A',f'' may be found by transferring the points g' and f'' horizontally to meet the slant side C,A', in n and o, on the outside of the cone. From the points n, o, on the slant, swing arcs into the pattern. It remains now to draw in the pattern itself, which should be easily followed from the diagram, Fig. 108.

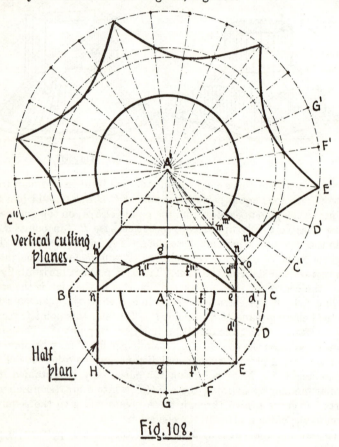

Fig. 108.

RECTANGULAR PIPE ON CONIC BASE

The illustration in Fig. 107 shows the reverse construction to that above. Instead of the cone resting on the square or rectangular base, the square or rectangular pipe rests on the conic base. This example, like the one above, is a straightforward case of conic intersection, and while it is not, perhaps, of common occurrence in practice, it is worthy of attention as a link in the chain of progressive studies. The perspective view in Fig. 107 shows the rectangular pipe placed on centre with the axis of the cone, but the example given for development in Fig. 109 presents the rectangular pipe off-centre both ways. The former case is, of course, the simpler, and if the directions for the latter are carefully followed no difficulty should be experienced with the other, or similar cases.

In setting out the plan and elevation to obtain the hyperbolic intersection curves, first draw the circle in plan representing the base of the cone, and then the rectangle in its required position. Next draw the elevation of the full cone, as at A',B',H'. Divide the base circle in the plan into twelve equal parts and letter them as shown. Project these points vertically upwards to the base of the cone in the elevation. Join the points on the base to the apex A'. Also, join the points on the circle to the centre apex A. Now, since these lines in the plan from the circle to the apex A are the same as those in the elevation from the base to the apex A', the points where the lines in the plan cross the rectangle may be located on those in the elevation by projecting the points of intersection vertically upward to the corresponding lines. Thus, the line D,A in the plan cuts the rectangle in point 4. This point projected vertically upwards will meet the line d,A' in the elevation in point $4''$. Similar points on the hyperbolic curves may be found for points 6,7,10,11,13,14. It will be seen that two of the four corner points, 7 and 14, fall on lines G,A and M,A respectively, but the other two corner points, 3 and 9, fall between two lines. In order to locate these points in the elevation, extra lines must be drawn from the centre apex A, through these points to the circle in the plan, and the points on the circle projected upwards to the base in the elevation, and from there to the apex A'. Then, points 3 and 9 projected vertically upwards to meet these lines in the elevation will give their respective positions on the hyperbolic curves. It will be observed that points 1, 2, and 8 fall on the outside verticals of the rectangular pipe in the elevation. There are two other points remaining, 5 and 12, which fall on the centreline, and are not so readily located in the elevation. To find the position of these two points, from the centre apex A in the plan, swing them on to the horizontal centreline B,H, and from there

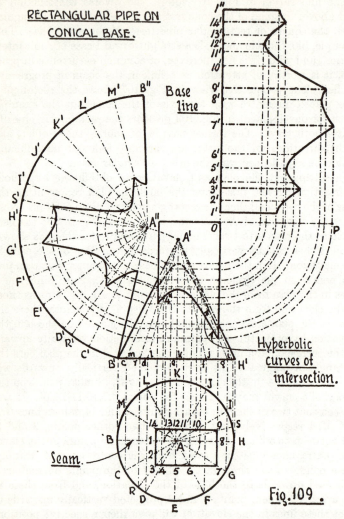

RECTANGULAR PIPE ON CONICAL BASE.

Base line

Hyperbolic curves of intersection.

Seam.

Fig. 109.

project them vertically upwards to the slant sides of the cone in the elevation. Then take them horizontally to the centreline, and each will give the highest point respectively on the front and back hyperbolic curves.

To develop the pattern for the conic section, project all the points located on the curves of intersection horizontally to the outside slant B',A' of the cone. In the diagram, Fig. 109, this slant side and all the points on it are revolved about B' to the new position $B'A''$, in order to keep the pattern clear of the elevation. The point A'' is now used as the apex for the pattern. From A'' swing arcs into the pattern from all the points on B',A''. Around the base arc B',B'' mark off twelve distances equal to those round the circle in the plan, and letter the points B',C', . . . L',M',B''. Join all these points to the apex A''. Next mark off the two extra points R and S in their relative positions between C',D' and $H'I'$, respectively. Join these also to the apex A''. All is now ready for drawing the curves of intersection in the pattern. Assuming that the seam is to be on the line, $B,1$, the curves should be carefully plotted in accordance with the illustration at Fig. 109.

To develop the pattern for the rectangular pipe project all the points on the curves of intersection in the elevation horizontally to the side of the pipe. In the ordinary process of development the pattern for this pipe should be "unrolled" horizontally, or at right angles to its central axis, but in the illustration given it is turned through 90 degrees for the convenience of space and general arrangement. Thus, all the points projected to the side of the rectangular pipe are revolved about the point O through 90 degrees to the line O,P, and the base line is projected vertically upwards. On the base line mark off a number of spaces corresponding to those round the rectangle in the plan, beginning at point 1. Number the points on the base line $1',2',3'$, . . . $13',14',1''$. Draw lines from each of these points at right angles to the base line, and project lines upwards from all the points on O,P, parallel to the base line to meet those at right angles to it. The intersection curves in the pattern may now be plotted. Assuming that the seam is to be in line with that of the cone, the first point on the curve will be that corresponding to point 1 in the plan. If the location of the correct points in the pattern should be a little puzzling, they may be readily traced up from the plan in the following manner. With the point of a pencil, or any other instrument, begin at point 1 in the plan, proceed to B on the circle, vertically upwards to B' on the base, up the slant of the cone to the intersection line, across to the other side of the rectangular pipe and round the quadrant to the line O,P, then upwards until it meets the line out from point $1'$ on the base line. Next, begin at point 2 in the plan, proceed to C on the circle, vertically upwards to c on the base, up the elevation line to the intersection, across to the quadrant and round to the line O,P, then upwards until it meets the line out from point 2′ on the base line. Next, begin at point 3 in the plan, proceed to R on the circle, vertically upwards to r on the base, up the elevation line to the corner point of the intersection line, across to the quadrant and round to the line O,P, then upwards until it meets the line out from point 3′ on the base line. Next, begin at point 4 in the plan, proceed to D on the circle, vertically upwards to d on the base, up the elevation line to 4″ on the intersection curve, across to the quadrant and round to the line O,P, then upwards until it meets the line out from 4′ on the base line.

When this process is properly grasped, the path of each point can be rapidly traced by the eye to its correct position in the pattern.

SKILL IN PATTERN-DRAFTING

It is a fairly obvious fact that skill in the manipulation of the hammer can be an effective time saver, but it is not so readily seen that skill in the application of the rule and compasses can be a greater, and even more effective, means of economy. Pattern-drafting cannot be dispensed with. Whether it be the preparation

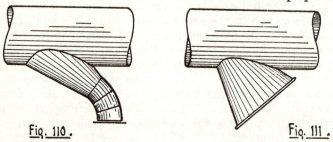

Fig. 110 .   Fig. 111 .

of templets to be used in mass production, or the drafting of a single pattern to be used once only, marking out has to be done, and this can be executed with speed and certainty by craftsmen well versed in methods of geometry. Mental haziness in this connection is responsible for more errors and undue delay than is generally realized, for no one can get on with his job if he does not know how. Skill in pattern-drafting is not difficult to acquire, if the subject is studied methodically. There are only three methods of development by means of which any pattern may be drafted, and those three methods, the Radial Line method, the Parallel Line method and Triangulation, form the basis of this course of study. The two chief sections into which the Radial Line method may be divided are problems involving the principles of the right cone and those of the oblique cone.

The two problems dealt with in Figs. 112 and 113, introduce a fresh point or principle in so far as the line of intersection in the elevation must be determined before the pattern can be developed. Both examples are of right conic frustums, and although that of Fig. 112 represents a conical spout fitting on a cylindrical body, as that of a pouring can, the same problem may be encountered in a variety of ways. The illustrations at Figs. 110 and 111 show two applications of the same problem, and although the cone in the latter case is reversed in position, the method of obtaining the line of intersection is precisely the same.

CONICAL SPOUT ON CYLINDRICAL BODY

The spout in Fig. 112 is shown proportionately larger than is usually required for a purpose of that kind. It is shown enlarged

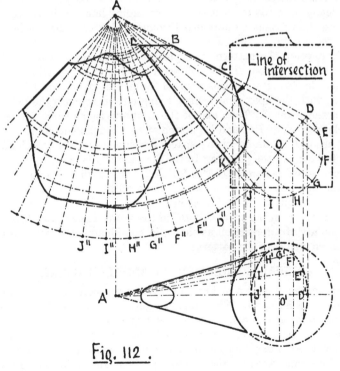

Fig. 112.

in order to emphasize the line of intersection and illustrate the processes of development more clearly.

In order to determine the line of intersection, first complete the elevation of the full cone ADJ by producing the sides BC and LK upwards to meet in the apex A, and downwards to any convenient position to form the base DJ. The distance AD must be made equal to AJ. On the base DJ describe a semicircle and divide it into six equal parts, as shown at D,E,F,G,H,I,J. From these points project lines perpendicularly back to the base DJ, and join the points obtained on the base to the apex A. Next project the base of the cone into the plan by dropping vertical lines from the points on DJ, to cross the horizontal centreline in the plan. Mark off, on the corresponding lines above and below the horizontal centreline $J'D'$, distances equal to those from the base DJ to the semicircle. Draw in the ellipse to represent the base, and join the points D',E',F',G',H',I',J', to the apex A'. From the points where these lines cross the circle representing the cylindrical body, project lines vertically upwards to meet the corresponding lines in the elevation. Points should thus be obtained through which to draw the line of intersection.

To develop the pattern, first project lines at right angles to the centreline AO, from all the points on the line of intersection to the slant side AJ. Repeat this with all the points on the top edge LB. Next, with the apex A as centre, swing out an arc into the pattern from the point J. Take one of the equal divisions from the semicircle, as DE, or EF, and beginning at any convenient point on the arc, as at D'', mark off twelve spaces which will represent the base of the full cone in the pattern. In the diagram at Fig. 112, six of these divisions are lettered $D'',E'',F'',G'',H'',I'',J''$. Join all these points to the apex A. Now describe arcs from all the points on the slant side AJ to cut the radial lines in the pattern. Assuming that the joint is to be on the short side BC of the spout, lines drawn through the points of intersection as shown in the illustration will give the required curves of the pattern.

CONICAL CONNECTION ON ANGULAR CORNER

The problem shown in Fig. 113 represents a conical spout on the corner of a hexagon. While this example is similar to the previous one in so far as it is a spout on a vessel, it also finds application as a conical connection on the corner of a square or rectangular duct. The method of development is precisely the same in either case, and the directions given for the previous problem shown in Fig. 112 will serve equally well for that of Fig. 113. The chief difference between these two problems will be seen in the form of the line of intersection in the elevation. The curve is more pronounced in the case of the angular corner, and from the point K, in Fig. 113, it dips downwards before rising to point C. This apparently simple variation is often

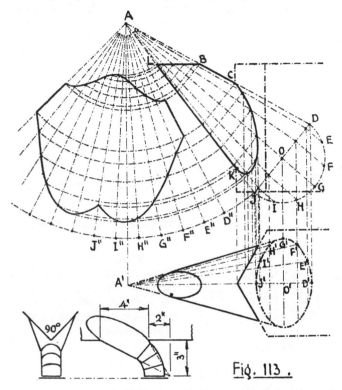

Fig. 113.

the cause of doubt and error, and the example is well worth the trouble of working in order to impress the difference on the memory.

The smaller diagram given at the bottom left-hand corner in Fig. 113 is offered as an additional example for practice.

BRANCH PIECES IN PIPE WORK

There is an extraordinary variety of designs possible for branch pieces in air duct work. Efficiency in the matter of air flow should be the guiding principle in designing a branch piece to satisfy any given set of conditions. The angle of divergence should be as small as possible, which means that the limbs forming the branches should not divide too abruptly like the arms of a Y-piece, but should diverge at an angle which allows the two holes or outlets at the double end to be as close together as practicability permits. The illustration at Fig. 114 shows a two-way branch piece with the holes at the top cut parallel with the hole at the base, and two full 90 degree bends leading from the branches. The fact that the holes at the top are parallel with the base allows the bends to be turned to any desired angle in the plan, as will be seen by that shown at *A*, without com-

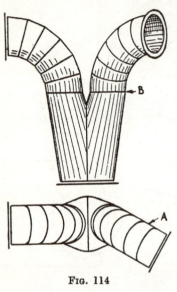

plicating the position of the bend. This condition is often very convenient when the exact angle cannot be predetermined, and the joint at *B* is left loose or unfixed for adjustment on site. The limbs of the branch piece form portions of oblique cones, and the method of developing the pattern is shown in Fig. 115.

TWO-WAY CONICAL BRANCH PIECE

The smaller figure in the bottom left-hand corner of the diagram at Fig. 115 represents a full plan and elevation of the two-way oblique conical branch piece which is shown developed above.

To draft the pattern, first draw the elevation of one limb, and

FIG. 114

produce the sides to obtain the full cone, as at 1,*A′*,7. Next draw the plan, with the apex *A* vertically below *A′*. Only a half-plan is given in the illustration, and is attached direct to the base of the elevation for convenience and simplicity in drafting the pattern. Divide the semicircle, representing the base of the cone, into six equal parts, and number the points from 1 to 7. With the apex *A* in the plan as centre, proceed in accordance with oblique cone practice, by swinging the plan lengths round from each point on the semi-circle to the base of the cone in the elevation. From the points on the base line 1,7, draw lines to the apex *A′*. These lines are TRUE-LENGTH lines. Next, with the apex *A′* as centre, swing arcs into the pattern from all the points on the base line, and also from all the points on the top edge of the cone where the true-length lines cross. Now take one of the equal divisions from the semicircle, as 1,2 or

2,3, and, starting from any convenient point on the outside curve in the pattern, as at 1', mark off twelve divisions by stepping over

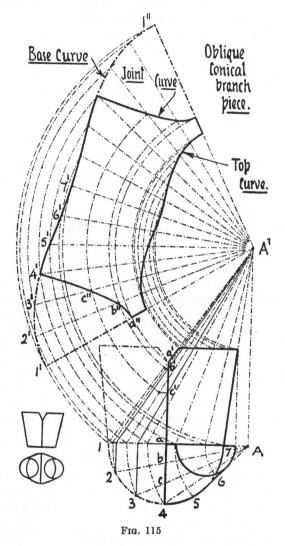

FIG. 115

from one curve to the next, as shown in the diagram. On reaching the inside curve from the base, as at 7', the spacing should proceed outwards to point 1″. A line drawn through these points should give the base curve in the pattern. To obtain the corresponding top curve, draw lines from all the points on the base curve to the apex A'. Where these radial lines cross the curves from the top edge of the cone points will be obtained through which the top curve may be drawn, as illustrated in the diagram.

Up to this point the procedure is precisely the same as for the full oblique cone frustum shown in Fig. 70, in the Second Course, except that the curves in the pattern are reversed in position in order to bring the seam on the opposite side. It remains now to obtain the curves which form the intersection or joint lines. In the plan it will be seen that the plan lengths $A,1$; $A,2$; and $A,3$, cross the joint line in points a,b,c. To obtain the true lengths of the portions A,a; A,b; and A,c, draw lines from points 2 and 3 perpendicular to the base line 1,7. It will be observed that the point 1 is already on the base line. From the points on the base, draw lines to the apex A' in the elevation. These lines are ELEVATION lines. From the points where the elevation lines cross the joint line, as at $a',b',$ and c', draw lines horizontally to meet the corresponding true length lines. For example, taking point 3 in the plan, the elevation line from that point crosses the joint line at c'. A short horizontal line drawn from point c' to meet the true length line from point 3 will give the required true length of A,c, from the apex A'. To obtain the true length of A,b, repeat this process from point b'. The distance A',a', is already a true length. These true lengths should now be swung into the pattern from the apex A', and where the resulting arcs cross the corresponding radial lines in the pattern, points will be obtained through which to draw the joint curves. Thus, the plan line $A,3$ crosses the joint line in c, and the corresponding true-length arc from c' crosses the radial line $A',3'$ in c''. Similar points at b'' and a'' afford positions through which to draw the curve $a'',b'',c'',4'$.

MULTIPLE CONICAL BRANCH PIECE

In the case of multiple branch pieces of three, four, or more limbs all equal, it is important to arrange the plan so that one limb lies on the horizontal centerline. The reason for this is that the development of the pattern for one limb can best be done when the elevation of the cone shows its maximum slant, which means that, in the plan, the centreline through the apex should be horizontal. The small three-way branch piece shown in the bottom left-hand corner in Fig. 116 is arranged so that the centreline of the limb marked B is on the horizontal centreline, and the pattern development is shown in the larger diagram above. The chief difference between this problem and that of the two-way branch piece in the previous example lies in the determination of the joint line in the

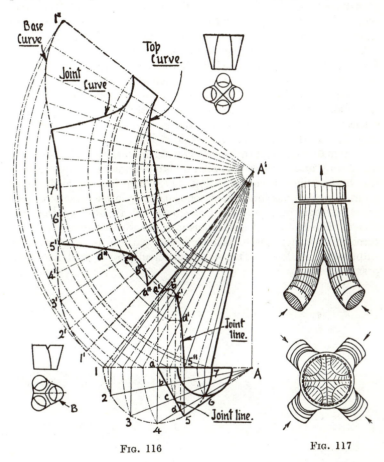

FIG. 116   FIG. 117

elevation. The joint line in the plan will occupy the position from the centre-point a to point 5. In the elevation this joint line is seen as the curve shown immediately above. To obtain this curve, divide the semicircle in the plan into six equal parts and number them from 1 to 7. Join these points to the apex A. From the points 1 to 7 on the semicircle draw perpendicular lines to the base line in the elevation, and join the points on the base line to the apex A'. These are the elevation lines corresponding to the plan lines $A,1$; $A,2$; . . . $A,7$. It will be seen that the plan lines cross the joint line in points a,b,c,d, with point 5 at the extremity. These points transferred vertically upwards to the corresponding elevation lines will give points a',b',c',d', with point $5''$ on the base line. A curve drawn through these points will give the joint line required.

The method of developing the pattern is the same as for that of the previous problem. Having obtained the pattern for the full cone, the joint curves may be obtained by projecting the points b',c',d' horizontally to the true length lines, and swinging arcs from the points obtained, and also from a' into the pattern to meet their respective true length lines from the base curve. The joint curves may then be drawn as shown at $a'',b'',c'',d'',5'$.

The smaller figure at the top right-hand corner, Fig.116, represents a four-way branch piece of similar design. A practical application of this branch piece is shown in Fig. 117, where four equal diameter pipes from four machines meet concentrically in one junction, from which a single pipe proceeds upwards to an air filter on the floor above. The method of development is exactly the same as for the three-way branch piece, although the pattern is a little more emphasized in form. This example should serve well for extra practice.

CHAPTER 9

THE PARALLEL LINE METHOD

THE grouping of similarities in methods of surface development simplifies the work and study of pattern-drafting. The parallel line method of development in the First Course was introduced with problems of tees and joints between pipes of equal diameter. In the Second Course, problems of tees and branches between pipes of unequal diameters marked a step forward. Now, in the Third Course, joints and branches on lobster-back bends made by pipes of different diameters present a further development in this type of problem.

It sometimes happens that a branch connection has to be made on the heel of an existing bend, similar to the example shown in Fig. 119. In dust-handling systems it is often advisable to fit a short stump, with a removable cap, on the heel of a bend to facilitate cleaning out in either direction by inserting brushes at the stump.

CONNECTION TO LOBSTER-BACK

The example shown at Fig. 118 may be regarded as a standard type of branch-to-bend problem, which is similar to that represented in the smaller diagram at the bottom left-hand corner, except that the larger pipe forms a bend. Before the patterns can be drafted, the correct determination of the joint line is necessary, and this may be done by following one or two simple principles. In the illustration, a half end elevation is given, in which the smaller semicircle represents a section of the cylindrical branch pipe, and the larger semicircle the end of the bend.

Divide the bottom quadrant of the smaller semicircle into three equal parts, as at 1,2,3,4. The middle point 4 on the smaller semicircle represents the outermost point of the cylinder which penetrates the bend. A line drawn vertically through that point, as at 4'4', may be considered as lying on the surface of the bend, and may be transferred to the front elevation by projecting the bottom point 4' horizontally to the edge of the bend, and then swinging it round from the centre 0. Where this curve crosses the centreline 4,4 of the branch pipe, the middle point d will be obtained on the joint line. To obtain other points on the joint line, draw lines vertically through points 2 and 3 on the smaller semicircle to meet the larger semi-circle in points 2' and 3'. From the latter points, project lines

151

horizontally to the edge of the bend, and from there swing curves round from the centre 0. Next, in the front elevation, describe the semicircle on the other end of the branch pipe, divide it into six equal

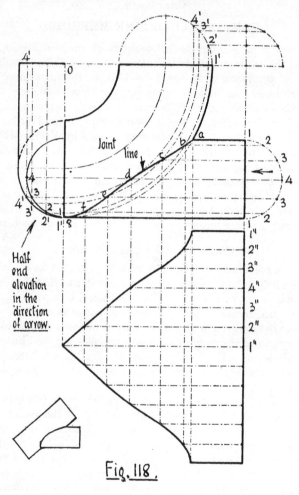

Fig. 118.

parts and number them 1,2,3,4,3,2,1, as shown in the diagram. This semicircle will be similar to the smaller one in the end elevation. Now, from the points on the second semicircle draw lines horizontally to meet the curves obtained from the corresponding points in the end elevation. The joint line may now be drawn in through points

a,b,c,d,e,f,g. The pattern for the branch pipe may now be "unrolled" in the usual way as shown in the illustration.

An important alternative method of obtaining the curves in the front elevation is shown at the top of the bend. The radius of the

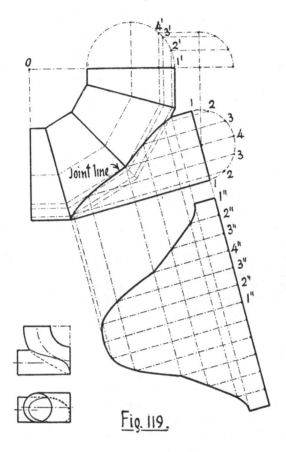

Fig. 119.

smaller quadrant corresponds to that of the branch pipe. This quadrant is divided into three equal parts and the points projected horizontally to meet the larger quadrant of the bend. From the point on the larger quadrant, lines are dropped perpendicularly to the edge of the bend, where, it will be seen, they meet the curves already obtained from the end elevation. This method of locating points from which to draw the curves is a little clearer than that shown in the end elevation, but, as the quadrants do not represent any exact view of the pipes, the method is conventional, and should be applied with caution to any new phase of the problem.

BRANCH ON FOUR-PIECE LOBSTER-BACK

To develop the pattern for the branch piece on the lobster-back bend shown in Fig. 119, first set out the elevation with the branch set at the required angle. Describe a semicircle on the top edge of the bend. Produce the edge, and on it describe a quadrant with a radius equal to that of the branch pipe. Divide the quadrant into three equal parts and project the points horizontally back to the semicircle. Number the points on the semicircle 1′,2′,3′,4′, and from them drop perpendicular lines to the edge of the bend. Continue these lines round the bend by drawing them parallel to the centre-line of each segment. Next, describe a semicircle on the end of the branch pipe, divide it into six equal parts and number the points 1,2,3,4,3,2,1. Project these points perpendicularly back to the edge of the pipe, and continue the lines onward, parallel to the central axis. Where these intersect the corresponding lines drawn round the bend sufficient points will be obtained through which to draw the joint line. The pattern may be "unrolled" in the usual way as shown in the illustration.

The smaller figure at the bottom left-hand corner represents a similar problem with the branch entering at the side. This is left as an example for further practice.

CLEANING OUT STUMP ON A LOBSTER-BACK

The problem given in Fig. 120 represents a short cylindrical stump on the heel of a bend. It is fitted with a removable cap to facilitate inspection or cleaning out. The pattern should be readily obtained by following the directions given for the previous problem. In the illustration the pattern is shown detached and not "unrolled" as in the previous examples.

The example shown in Fig. 121 represents a joint between a smooth bend and a pipe of equal diameter. In this case the joint line may be located by describing semicircles on the ends of cross-sections of the pipes, dividing the near-side quadrants into three equal parts, projecting the points square on to the pipe edges, and producing the lines on the pipe surfaces until they meet at the joint line.

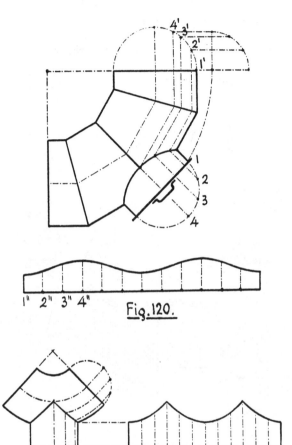

4' 3'
2'
1'

1
2
3
4

1" 2" 3" 4"

Fig. 120.

Fig. 121.

LOBSTER-BACK BRANCH

The problem shown in Fig. 122 is, in effect, the reverse of those given in Figs. 119 and 120, inasmuch as the diameter of the bend is smaller than that of the cylindrical pipe, which gives the bend the

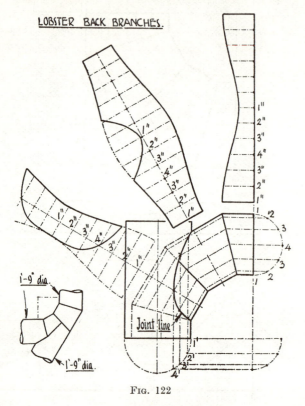

LOBSTER BACK BRANCHES.

FIG. 122

position of branch instead of main. Before the patterns for the segments of the bend can be obtained, the joint line must be correctly determined.

First set out the full elevation of the main cylindrical pipe and the lobster-back bend, and on the base of the main pipe describe a semicircle. Extend the base of the main pipe, and on it describe a quadrant with a radius equal to that of the lobster-back pipe. Divide the quadrant into three equal parts and project the points horizontally to meet the semicircle in points 1′,2′,3′,4′. From these points draw lines perpendicularly to the base of the main pipe. Now, instead of swinging them round the bend, as in the previous examples, draw lines upwards on the surface of the main pipe parallel to its centreline. Next describe a semicircle on the end of the bend, divide it into six equal parts and number them 1,2,3,4,3,2,1. Project these points back perpendicularly to the end of the pipe, and produce them round the bend in lines parallel to the centrelines of the segments. Where these lines meet the corresponding ones from the base of the main pipe, points will be afforded through which to draw the joint curve as illustrated in the diagram. The patterns for the segments may now be developed in accordance with the method given in Fig. 36.

The smaller figure in the bottom left-hand corner is another example of a lobster-back branch, in this case between pipes of equal diameters. This problem is given for additional practice.

AUXILIARY PROJECTIONS

The drafting of a pattern is usually a simple process once the principles of the Radial Line method, the Parallel Line method, and Triangulation have been properly grasped. It is the preliminary operations in preparing the views or conditions necessary for the development which often prove intricate or puzzling. For example, the open chute shown in the top left-hand corner in Fig. 125 is a straightforward problem of the oblique cylinder, and was dealt with in the First Course in Fig. 38. The example shown at (b), Fig. 125, is the same type of chute, except that it inclines at an angle sideways. This problem, which is often met with in practice, is not so simple as its predecessor. The sideways angle makes all the difference.

In each of the problems shown in Figs. 123, 124, and 125 the principles of auxiliary projection, which were introduced in the Second Course, will be needed to produce the necessary views for developing the patterns. It may be remembered that most auxiliary projections can become ordinary plan views if the drawing be turned round to suit that position Thus, in Fig. 123, the projected view may be regarded in that light, as also may that in Fig. 124, if the illustrations be turned so that the projections come vertically below the elevations.

RECTANGULAR BRANCH ON CYLINDRICAL PIPE

The problem shown in Fig. 123 represents a horizontal rectangular pipe entering a cylindrical pipe inclined at an angle of nearly 45 degrees. The projected view should be obtained looking in the direction of the arrow. It will be seen that the width of the rectangular pipe is just one-half of the diameter of the cylinder. For convenience, the corners of the open end of the rectangular pipe are lettered A,B,C,D, and the corresponding corners at the other end are

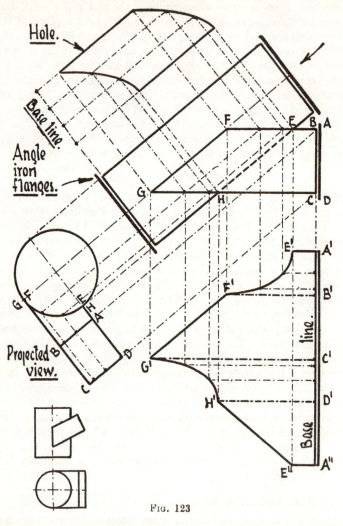

Fig. 123

lettered E,F,G,H. The top and bottom plates $ABFE$ and $CDHG$ intersect the cylinder in similar quadrants, which cut obliquely across the pipe, as at EF and GH. The true shape of the curve at these positions will be elliptical.

To "unroll" the pattern, produce the base of the rectangular pipe, and space off the distances A',B' ; B',C' ; C',D', and D',A'', equal to the true distances round the rectangle. The distances A',B' and C',D' should be obtained direct from the projected view, and B',C' and D',A'' from the elevation. Next draw horizontal lines from these points at right angles to the base line, and then from the corner points E,F,G,H in the elevation drop vertical lines into the pattern to meet the horizontal lines in E',F',G',H',E''. Join A',E' ; F',G' ; H',E'' ; E'',A''. It remains now to determine the curves E',F' and G',H'. Divide the quadrant EF in the projected view into three equal parts, and from the points of division draw lines parallel to the sides of the rectangular pipe to cut AB and CD into three parts, but not equal, as will be seen in the diagram. In the pattern, divide the spaces A',B', and C',D', each into three parts corresponding to those on AB and CD, and from these points draw horizontal lines into the pattern at right angles to the base line. Now, from the points on the quadrant in the projected view, project lines into the elevation to intersect GH and EF. From the points obtained on GH and EF, drop vertical lines into the pattern to meet the corresponding horizontal lines from the base line. Points will be thus afforded through which to draw the elliptical curves from E' to F' and G' to H'. To set out the shape of the hole in the cylindrical pipe, first observe that it extends round the girth of the pipe a distance equal to the curve of the quadrant in the projected view. Now project a base line at right angles to the central axis of the cylinder, as shown in the diagram, from the base of the pipe. On this base line mark off three divisions equal to those round the quadrant, and project lines from these points parallel to the central axis of the cylinder. Next, from the points on EF and GH project lines at right angles to the central axis to meet those from the base line. Points should thus be obtained through which to draw the shape of the hole, as illustrated in the diagram.

CROSS PIPES WHICH PARTIALLY INTERSECT

It is the unusual problem which often proves difficult for the pattern-drafter, not always because it is perplexing, but rather because it is different, perhaps in some minor detail, from the usual mode of application. One would hardly expect to find a cross piece, such as that shown in Fig. 124, in general use, and certainly not in duct work. Nevertheless, it finds useful application in some arrangements of feed pipes to machines dealing with grain or coarse powder, where the cross pipe may form an effective overflow unit. In some respects it is similar to the previous example. The chief difference is that the intersection is only partial instead of full.

First, set out the elevation showing the pipes crossing at the required angle. Next obtain the projected view with the pipe centres

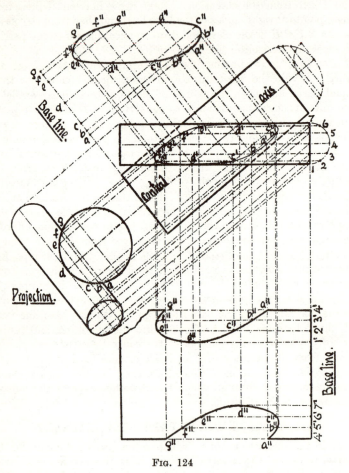

FIG. 124

set at the required distance apart. In this case the larger pipe cuts into the smaller pipe as far as the centreline. The ellipse representing the end of the smaller pipe in the projected view is obtained in the usual way by describing a semicircle on the end of the pipe in the elevation, dividing it into six equal parts, projecting lines from these points perpendicularly back to the end of the pipe, and from the points on the end of the pipe, drawing lines into the projected view to obtain those points on the ellipse. From the points on the ellipse draw lines parallel to the centreline of the smaller pipe to cut the circle of the larger pipe in points a,b,c,b,e,f,g. From these points project lines into the elevation. Now, from points 1 to 7 on the semi-circle, draw lines parallel to the centreline of the smaller pipe to meet the corresponding lines projected from the points a to g. Thus, the centre point 4 on the semicircle may be traced back to points a and g in the projected view. Therefore the horizontal line from point 4 meets the projected lines from a and g in points a' and g'. In this way all the points on the line of intersection in the elevation may be located, and the curve drawn in.

To "unroll" the pattern for the smaller pipe, produce the base line from the end of the pipe, and mark off twelve spaces equal to those round the semicircle. Assuming that the seam is to be on the inside, or on line $4,a',g'$, passing through the middle of the intersection, the line $4,a',g'$ will form the ends of the pattern. Therefore, on the base line, number the three spaces at each end so that the point $4'$ falls on the outside, as at $1',2',3',4'$, and $4',5',6',7'$. Next draw lines from these points at right angles to the base line, and from the points on the line of intersection in the elevation, drop vertical lines into the pattern to meet the corresponding lines from the base line in points $a'',b'',c'',d'',e'',f'',g''$.

To develop the shape of the hole in the larger pipe, extend a base line from the bottom edge at right angles to its central axis, and mark off the distances a,b,c,d,e,f,g, equal to those round the curve in the projected view. From these points project lines at right angles to the base line. Next, from all the points round the line of intersection in the elevation draw lines into the pattern parallel to the base line, to meet the corresponding lines projected at right angles. Thus, the line projected from the middle point, d, meets the two lines drawn from d',d' in points d'',d''. In this way sufficient points should be obtained through which to draw the required shape of the hole.

It will be seen that a semicircle is described on the top end of the larger pipe, and a quadrant of the same radius as that of the smaller pipe is described on its centreline. The quadrant is divided into three equal parts, and lines are drawn through the points parallel to the end of the pipe. These lines intersect the semicircle in positions similar to those at a,b,c,d,e,f,g, in the projected view, and if lines were drawn from these positions parallel to the central axis of the pipe, the same points on the line of intersection would be obtained.

By the use of this semicircle and quadrant the work involved in obtaining the projection might be cut out, since that view would not be needed. However, it is not always wise to introduce a conventional method such as this, unless its mode of application can be clearly seen and understood. It is often better to follow well-defined principles, even involving a little more work, than to risk uncertain short cuts which might lead to confusion.

AN OBLIQUE CYLINDRICAL CHUTE

The oblique cylindrical chute shown at Fig. 125 is one which is often used on a circular outlet from a mixer, or on the end of a screw

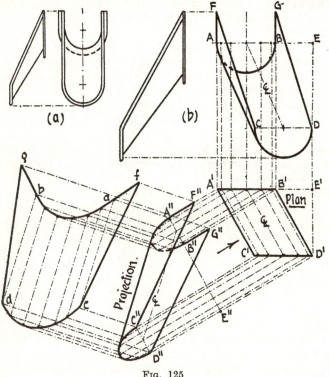

Fig. 125

conveyer. When the point of delivery is on centre with the outlet, the type shown at (a) will be required. When it is necessary for the point of delivery to be off-centre with the outlet, the type shown at (b) will be needed. Before the pattern for the latter can be drafted a projected view must be obtained in which the chute lies flat, so to speak, on the paper. This view, in the illustration given, is in the direction of the arrow, but placed below the plan instead of above for the convenience of suitable arrangement. To obtain this projection, divide the top semicircle in the elevation into six equal parts, and through the points draw vertical lines to the horizontal centreline AB above and to the corresponding plan line $A'B'$ below. From $A'B'$ draw lines parallel to the centreline in the plan to $C'D'$ at the other end. Next project all the points on $A'B'$ and $C'D'$ at right angles to the centreline into the projection. Draw the line $A''E''$ at right angles to the projected lines, and mark off $E''D''$ equal in length to the vertical height ED in the elevation. Draw $C''D''$ parallel to $A''E''$. Now, on $A''B''$ and $C''D''$, describe the semi-ellipses, making the depths on the projected lines equal to those in the semicircle on AB in the elevation. Next mark off $A''F''$ and $B''G''$ equal to AF and BG, which should complete all the points in the projected view.

To "unroll" the pattern, project all the points at right angles to the centreline CL of the projection. Then, take the true distance, FA, from the elevation, and, from any convenient point f on the projected line from F'', space off the distance, fa, by stepping over from f to the line projected from A''. Follow this up by taking one of the six equal divisions from the semicircle AB in the elevation, and from point a in the pattern mark off the next six spaces to b by stepping over from one line to the next, being careful each time to see that the "next" line is the one projected from the "next" point on the ellipse. Now take the true distance BG from the elevation and mark off bg in the pattern, to finish on the line projected from G'' in the projection. Draw in the contour from f to g. The bottom contour from c to d may be obtained by drawing lines from all the points on the curve a to b parallel to the centreline of the projection. Where these lines meet the corresponding lines from the bottom ellipse, points will be afforded through which to draw in the curve. Join gd and fc. This completes the pattern.

ORNAMENTAL BOWLS AND VASES

Segmental mouldings form an attractive section of developments by the Parallel Line method. Whether the subject be curbs, downspout heads, roof finials or ornamental bowls and vases, artistic treatment is an outstanding feature of the designs. The curves outlined in the elevation are generally intended to form pleasing contours, and in addition, in the case of bowls and vases, decorative tooling and chasing may still further enhance the appearance. Those problems of mouldings which have already been considered in the First and Second Courses required only one contour for the patterns. Often, where the segments were all alike, the same contour repeated left- and right-hand gave a pattern symmetrical about a centreline.

The illustration at Fig. 126 represents a twelve-sided plant vase in which all the segments are similar, thus requiring one pattern only. This diagram also illustrates the effect of ordinary chasing and hammer-marking as a form of decoration. The problems presented in this course each require three different patterns, since the segments *A,B,C*, as shown in the plan at Fig. 127, are different.

SEGMENTAL ROSE BOWL

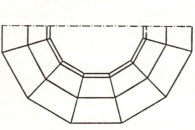

Fiq. 126.

To develop the pattern for the segment at *A*, Fig. 127, the actual length of its centre-line 1″,8″ must first be set out. This may be obtained from the outside contour 1′,8′ in the elevation. First divide the contour into a number of parts, as at 1′,2′,3′,4′,5′,6′,7′,8′, according to convenience. The divisions need not be equal. From these points drop vertical lines into the plan to cut across the segment *A*. Extend the centreline of segment *A* into the pattern, and on the extension mark off a number of spaces equal and corresponding to those, 1′ to 8′, marked on the contour in the elevation. Through the points thus marked on the centreline in the pattern, draw lines at right angles to the centreline. Next, from the points where the vertical lines from 1′ to 8′ cut across the sides of segment *A* in the plan, project lines into the pattern parallel to the centreline. Where these lines meet the corresponding cross lines at right angles, points will be afforded through which to draw the side contours of the pattern.

Now, regarding segment *B*, in the plan it will readily be seen that the width from 1 to 5 is less than that from 1 to 5 in segment *A*.

Therefore, before the pattern for segment B can be developed it will be necessary to obtain a true contour of a cross-section at right angles to its axis. This contour in the illustration is projected in

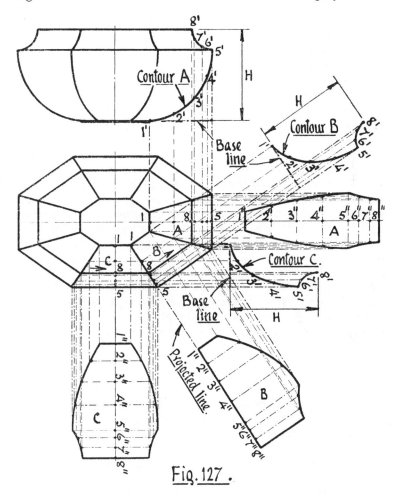

Fig. 127.

the direction of the arrow, parallel to its axis. To determine the true shape of this contour, through all those points which occur on the joint line between A and B, draw lines parallel to the outside line of segment B, in other words parallel to its axis. Extend these lines backward to cut the joint line between B and C, and forward to the position of contour B. Draw a base line in any convenient position across these lines at right angles to them. Let this correspond to the base line below contour A in the elevation, then the height of each point above the base line will be the same in each contour. Therefore, take the vertical heights of points $2',3',4',5',6',7'8,'$ from the base line in the elevation and mark these distances off on the corresponding lines above the base line in the projected contour. A curve drawn through these points will give the true form of a cross-section through segment B at right angles to its axis. This contour is similar in form to contour A, but, it will be observed, while the height H is the same in each case, the width is less in contour B. Therefore, the actual length of the contour B will be shorter than that of A.

Now, to develop the pattern, project any line at right angles to the axis of the segment. In this case the joint line between segments B and C is at right angles to the axis of segment B, therefore the required line is produced from that joint line. On the projected line mark off the distances $1'',2'',3'',4'',5'',6'',7'',8''$ equal to the spacings $1',2',3',4',5',6',7',8'$ from contour B. Through the points thus marked off, draw lines at right angles to the projected line. Next, from the points on the joint line between segments A and B, draw lines into the pattern parallel to the projected line. Where these lines meet the corresponding lines at right angles, points will be afforded through which to draw the joint curve in the pattern.

Again, regarding segment C, in the plan it will be seen that the width from 1 to 5 is still less than that from 1 to 5 in segment B, and, therefore, before the pattern can be developed, it is essential to project a true cross-sectional contour at right angles to its axis. To obtain this, through all the points which occur on the joint line between segments B and C, draw lines parallel to the outside line of segment C. Extend these lines backward to the opposite joint line on segment C, and forward to the position of contour C. Draw a base line in any convenient position across these lines at right angles to them, and again let this correspond to the base line below contour A in the elevation. Then the height of each point above the base line will be the same in each contour. It should be observed, however, that the term "above," in the case of contour C, means perpendicular to the base line, even though the "heights" are actually horizontal. Take the vertical heights of points $2',3',4,'5,'6',7',8'$ from the base line in the elevation, and mark these distances off on the corresponding lines above the base line in the projected contour. A curve drawn through these points will give the required true form of the contour. The actual length of this contour is still shorter than that of B. Now, to develop the pattern for segment C, project the centreline 1,5 into the pattern and mark off a number of spaces 1″,2″,3″,4″,5″,6″,7″,8″ equal to those around contour C. Through these points draw lines at right angles to the centreline. Next, from the points on the joint lines on each side of segment, C, draw lines into the pattern parallel to the centreline. Where these lines meet the corresponding lines at right angles, points will be afforded through which to draw the joint curves of the pattern.

SEGMENTAL VASE

The ornamental vase in Fig. 128 presents a similar problem of development to the rose bowl in the previous example. It will be seen that the plan is similar in form, but the elevation is quite different. The patterns for the segments A, B, and C are obtained in precisely the same way, and, for this reason, only the middle one, B, is developed in this example. There is one point of difference in the plan of segment B, which makes the pattern for this worthy of further attention. In the previous example, the joint line between segments B and C is at right angles to the outside line, or axis, of segment B. This produces the straight line 1″,8″ on one side of the pattern. In this example, Fig. 128, the joint line is not only other than a right angle, but leans in the same way, or on the same side, as that of the opposite joint line between A and B. This produces contours on both sides of the pattern which curve in similar directions instead of opposite, as they do in the patterns for A and C.

To obtain the projected contour for segment B, number the points in the elevation as shown at 1,2,3,4,5,6,7,8. In this example for the sake of simplicity only the prominent points of the contour are numbered, but it would tend to give a truer curve in the projected contour if more divisions were used. From the numbered points in the elevation drop vertical lines into the plan to cut the joint line between segments A and B. From the points obtained on this joint line project lines parallel to the outside line, or axis, of segment B, producing them to the position of contour B. Draw the base line at right angles to the projected lines. Next, take the vertical heights of the points 2,3,4,5,6,7,8 in the elevation and mark them off on the corresponding lines above the base line in the projected contour. Thus, points will be obtained through which to draw the curve of contour B.

To develop the pattern for segment B, project a line at right angles to the segment, and on it mark off a number of spaces equal to those around the contour B, as shown at 1″,2″,3″,4″,5″,6″,7″,8″. From these points draw lines at right angles to the projected line. Next, from all the points on the joint lines on both sides of segment *B*, draw lines into the pattern parallel to the projected line. Where these

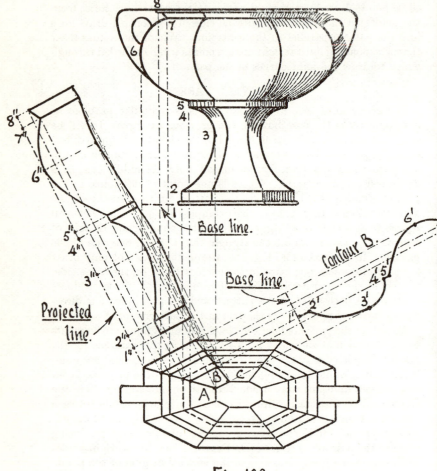

Fig. 128.

lines meet the corresponding lines at right angles, points will be afforded through which to draw the joint curves in the pattern.

The development of the remaining two segments, *A* and *C*, should present no difficulty if these principles have been fully grasped.

FURTHER PROBLEMS OF THE CENTRAL SPHERE

There is nothing to be gained by making a problem appear more complicated than it need be, yet simplicity is not always obtained by cutting and trimming. Abbreviation should be regarded with care. Short cut methods are often more difficult to understand than fully detailed examples, particularly when the former assume a knowledge of principles somewhat outside the scope of the problem. On the other hand, it is reasonable to take it for granted that certain methods are understood when they have been fully discussed in previous problems. For instance, in Fig. 131, the patterns shown developed are of cylindrical and right conic frustums, and since these have been dealt with at some length in earlier chapters, it should be unnecessary to explain the pattern developments in full detail. The chief points of interest in these examples are the lines of intersection between the parts.

The problems presented in Figs. 129 and 132A are further examples of intersections between cones and cylinders around an imaginary

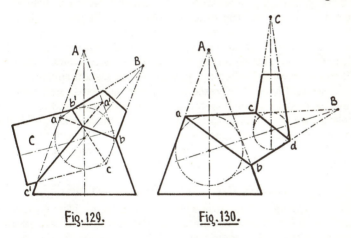

Fig. 129.   Fig. 130.

central sphere. This type of intersection has many interesting applications, not only in pipe work but also in the form of hoppers, hoods, and chutes. The example given in Fig. 132 represents a typical chute used for liquid. The particular point of importance in connection with this class of problem is that the lines of intersection in the elevation are straight, which makes the solution much more simple than those involving curved lines of intersection. Fig. 130 shows three right cones intersecting around central spheres.

To set out the elevation of those cones, first set out the centrelines in their required positions; then draw the large cone A and, on the point where the two centrelines cross, describe the first circle with a radius which just touches the sides of the cone. Next mark off the apex of the second cone B, and draw the sides of that cone so that they also just touch the first circle. Then the joint line between those two cones is a straight line between the opposite points of intersection, as at a,b. Now, on the point where the second and third centrelines cross, describe the second circle with a radius which just touches the sides of the middle cone B. Next, mark off the apex of the third cone, C, and draw the sides of that cone so that they just touch the second circle. The joint line between those two cones will then be the straight line, c,b, between the opposite points of intersection.

The example shown at Fig. 129 represents the intersection of two cones and a cylinder around one central sphere. Again, the centrelines should first be drawn in their respective positions, in this case all passing through one centre. Next draw the elevation of the large cone A, and describe the central circle so that it just touches the sides of the cone. Now mark off the apex of the second cone B, and draw its sides so that they also just touch the central sphere. The sides of the cylinder should next be drawn parallel to its centreline and so far apart that they too just touch the central sphere.

The straight lines of intersection may now be located, first between A and B, next between B and C, and then between C and A, respectively ab, $b'c$, $c'a'$. These three lines all cross at the same point, but that point is not at the centre of the sphere. The portion of each joint line which forms a part of the composite intersection is shown in full line, while the remainder is dotted.

TRIPLE INTERSECTION DEVELOPED

Now referring to Fig. 131, which is a similar problem to that of Fig. 129, the patterns for the cones and the cylinder are treated independently. The semicircle on the base of the larger cone is divided into six equal parts. The base curve in the pattern, with the apex A as centre, is swung out from the base of the cone and twelve divisions equal to those on the semicircle are marked off round the curve. These points are now joined to the apex A. The points on the semicircle are projected perpendicularly up to the base of the cone, and from the base the points are joined to the apex. Where the lines cross the intersection, the points are projected horizontally to the outside slant of the cone, and from the outside slant the points are swung into the pattern to meet the corresponding lines from the base curve. Thus, points are obtained through which to draw the joint curve in the pattern.

To obtain the pattern for the smaller cone, a base must first be

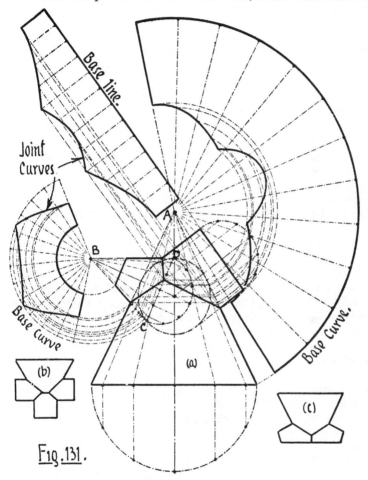

Fig. 131.

drawn at right angles to its centreline, as, in this case, at CD, which cuts through the point where the lines of intersection meet. The semicircle described on this base is divided into six equal parts, and the points projected perpendicularly back to the base. From the base the points are joined to the apex B, and, where the lines cross the intersection, those points are projected at right angles to the centreline, or parallel to the base, to the outside slant of the cone. Then, from the apex B, the points on the slant are swung into the pattern. The base curve in the pattern, with the apex B as centre, is swung out from the base CD, and twelve divisions, equal to those on the semicircle on CD, are marked off round the curve. These points are now joined to the apex B, and where they cross the curves from the slant of the cone, points will be afforded through which to draw the joint curves as shown in the diagram.

To "unroll" the pattern for the cylindrical connection the semicircle described on the end of it is divided into six equal parts. The points are then projected perpendicularly back to the end, and onward to cut the intersection lines. The base line in the pattern is extended from the end, or base of the cylinder, and twelve divisions marked on it equal to those on the semicircle. From these points, lines are drawn at right angles to the base line. Next, from the points where the lines on the cylinder cut the intersection, more lines are projected into the pattern parallel to the base line to meet those drawn at right angles to it. Points should thus be afforded through which to draw the joint curve. The examples illustrated at (b) and (c) should serve well for further practice.

AN OFF-CENTRE FUNNEL

The example shown at Fig. 132 is a typical form of an off-centre funnel used for liquid passing from one point to another, as from a storage tank to the filling inlet of a boiler or condenser. The middle conical portion may, of course, be longer or more horizontal to suit circumstances. In this particular construction it will be observed that one side of the top cone forms a straight line with the corresponding side of the middle cone. This introduces the application of a further principle in locating the point C on the straight line to which the intersection line should be joined. It has a definite position and only one spot will satisfy the correct solution of the problem. Any other point will result in the two ellipses between the cones not fitting correctly.

To set out the elevation of this problem, first mark off the base diameter EF, of the top of the cone, and draw the straight line at the required angle to form the side of the two cones. Next, draw the central axis of the top cone, locating the apex A, which is also on the straight line. The circle representing the central sphere may now be drawn, with a radius which will give the required size of the middle cone at that end. Next, the centreline of the cylindrical portion of

the chute should be drawn at the required off-centre position, and
the central sphere or circle described with its centre on the centreline

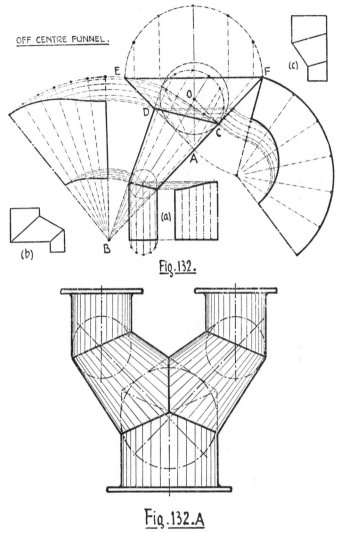

OFF CENTRE FUNNEL.

Fig. 132.

Fig. 132.A

of the cylinder in such a position that it touches the straight line
FB of the two cones. The second side of the middle cone may now
be drawn just touching the two circles and meeting the first side in

apex B. This side cuts the other side of the top cone in point D, which locates one end of the line of intersection. Now, to find point C, draw, from the centre O of the circle, a line at right angles to the straight side FAB. This will locate the point C, which is the other end of the line of intersection CD.

The development of the patterns may be followed from the directions given for the previous example, since these also are cylindrical and right conic frustums. The two smaller diagrams at (b) and (c) are given as additional examples for practice.

The illustration at Fig. 132a represents a "Y" piece constructed on these principles, and may be regarded as a double transition piece, or a right- and left-hand combination of a construction similar to that of Fig. 130. The examples discussed in this section are only a few of those which are possible and practicable in the application of this principle of intersections round a central sphere, and many more may be devised to suit given conditions or to meet special circumstances.

Chapter 10

TRIANGULATION

THIS is the Third Course of triangulation. As the work advances it is to be expected that the style of problem will emerge from the stage of simplicity into something requiring deeper thought and more careful attention. Nevertheless, the fundamental principles remain the same. The subsidiary principles, building up or adding to the work already done, form fresh view-points by means of which the resources of the pattern-drafter may be further increased.

The problems of triangulation of the First Course were those of transformers between two parallel planes, involving only one vertical height in the solution. Those of the Second Course were of transformers between two planes inclined at an angle to each other, requiring two or more vertical heights. In this course, transformers between curved, angular, or complex surfaces will be dealt with, involving special considerations in the methods of triangulating the surfaces.

OVAL-TAPERED CONNECTION TO CYLINDER

The connecting piece shown in Fig. 133 transforms from an oval hole on a curved surface to the elliptical cut of the pipe above. The smaller diagram in the bottom left-hand corner represents a typical example of the practical application of this kind of problem.

For the pattern-development, since the plan is symmetrical about the horizontal centreline, only one half will be triangulated. Divide the semicircle in the plan into six equal parts, and divide the corresponding half-oval into six equal parts. Triangulate the surface by drawing the zigzag line from 1 to 14, and number the points accordingly. Transfer these points vertically upwards to the base and the top, and join the corresponding points as in the plan. The actual shape of the hole in the top is an ellipse, and the true shape of this should be set out by projecting lines at right angles to the top edge from the points on it obtained from the plan. Then, cut off each of these lines equal in length to the corresponding width of the circle in the plan. The result should give sufficient points through which to draw the half-ellipse as shown in the illustration, numbered from 2′ to 14′. Next, erect a vertical height line, and on to it project all the points on the top and bottom edges in the elevation.

The first line in the pattern 1″,2″ may be taken direct from the elevation, since 1′,2′ is the required true length. Mark this off in any convenient position in the pattern. Next take the plan length 2,3, and mark it off at right angles to the vertical height along the base line from point 3′. Take the true length diagonal up to the top point

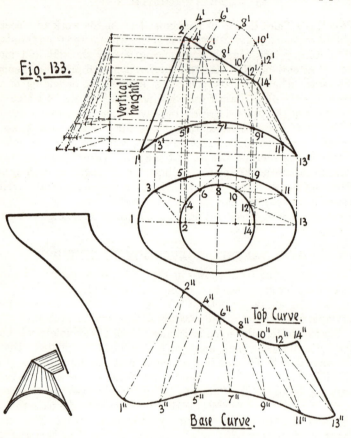

Fig. 133.

level with 2′, and from 2″ in the pattern swing off an arc through point 3″. The next line in the pattern, 1″,3″, marks an important step in the development. It will be seen that neither the distance 1,3 in the plan nor 1′,3′ in the elevation gives the true distance between those points. Therefore the rule of plan length at right angles to vertical height must be used in order to obtain the true length. Take the distance 1,3 from the plan and mark it off at right angles to the vertical height along the base line from 1'. Take the true length diagonal up to the point level with 3', and from point 1" in the pattern describe an arc cutting the previous arc in point 3".

For the second triangle take the plan length 3,4, and mark it off at right angles to the vertical height along the base line from point 3'. Take the true length diagonal up to the point level with 4', and from 3" in the pattern swing off an arc through point 4". The next true distance 2',4' should be taken direct from the semi-ellipse in the elevation, and from point 2" in the pattern, an arc described cutting the previous arc in point 4".

For the third triangle take the plan length 4,5, and mark this off at right angles to the vertical height along the base line from point 5'. Take the true length diagonal up to the point level with 4' and from point 4" in the pattern swing off an arc through point 5". Next take the plan length 3,5, and mark this off at right angles to the vertical height along the base line from point 3'. Take the true length diagonal up to the point level with 5', and from point 3" in the pattern describe an arc cutting the previous arc in 5".

For the fourth triangle repeat this process with the plan length 5,6, taking care to triangulate it against the vertical height between points 5' and 6'. The true distance 4",6" in the pattern is then obtained direct from the semi-ellipse in the elevation, as from 4',6'. The remainder of the pattern should be easily followed, since the process is the same to the end. The chief points to observe are, that the true distances 1",3" to 11",13" round the base curve must be obtained by triangulating the plan lengths against the respective vertical heights, and the true distances 2",4" to 12",14" round the top curve may be obtained direct from the semi-ellipse in the elevation.

CORNER WALL HOPPER

The pattern given in Fig. 134 represents a type of hopper suitable for the occupation of a corner of a room. The back of the hopper extends upwards to form a protective cover on the two sides of the wall, and the front of the hopper forms a quadrant, as shown in the plan. The hole at the back in the bottom gives access to a chute which passes through the floor, not shown, to convey the material to a receptacle below. In the plan it will be seen that the hopper is symmetrical about a line from the corner at the back to the centre at the front, as at 1,14. Assuming that the seam is to be at the front, from 13 to 14, it will be most convenient to divide the semicircle in the plan into eight equal parts instead of the usual six, and begin the triangulation and system of numbering at the back. Thus divide the surface into triangles and number the points as shown in the plan, from 1 to 14. Erect a vertical height line in the elevation,

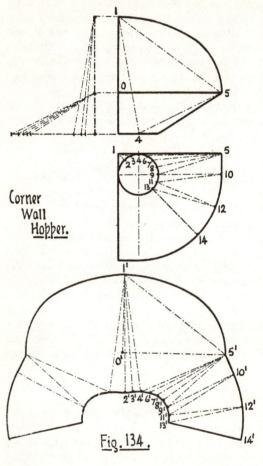

Corner Wall Hopper.

Fig. 134.

and project the two heights, the hopper front and the middle of the back, horizontally to it.

Then, to develop the pattern, take the plan length 1,2, and mark it off along the base line at right angles to the vertical height. Take the true length diagonal up to the top point and mark off 1',2' in the pattern. Next take the plan length 1,3, and mark it off along the base line at right angles to the vertical height. Take the true length diagonal up to the top point and from point 1' swing off an arc through point 3' in the pattern. Next take the true distance 2,3, direct from the plan and from point 2' in the pattern describe an arc cutting the previous arc in point 3'. For the second triangle take the plan length 1,4, and mark it off along the base line at right angles to the vertical height. Take the true length diagonal up to the top point, and from point 1' in the pattern, swing off an arc through point 4'. Next take the true distance 3,4 direct from the plan, and from point 3' in the pattern describe an arc cutting the previous arc in point 4'.

The next step is an important one. Point 5, as will be seen in the elevation, is the first point on the lower height line, level with the front of the hopper. The top edge of the hopper between 1 and 5 is a quarter of an oval with the centre of its axes at O, but for the purposes of development a straight line may be taken from 1 to 5. Then the triangle 1,4,5 is a flat piece of metal which may be triangulated into the pattern in the ordinary way, and the shape of the oval curve may be determined afterward. Since the triangle 1,4,5 is in a vertical plane, and lies flat against the wall, the sides 1,4; and 4,5, in the elevation are true lengths. Therefore, take the distance 1,5 direct from the elevation, and from point 1' in the pattern swing off an arc through point 5'. Next take 4,5 direct from the elevation, and from point 4' in the pattern describe an arc cutting the previous arc in point 5'. Join 1',4'; 1',5'; and 4',5'. The remainder of the pattern to point 14 is straightforward triangulation, and should be easily followed from this point. All the vertical heights from points 5 to 14 will take the lower height, level with point 5 in the elevation.

Now, to determine the oval curve in the pattern, since O, in the elevation, is the centre of the axes, that point may easily be determined in the pattern by taking the distance $O,5$ in the compasses, and from 5' in the pattern swinging an arc through point O'. Next, take $O,1$ from the elevation, and from 1' in the pattern describe an arc cutting the previous arc in O'. Join $O',1'$ and $O',5'$. Regarding these now as the semi-axes of the complete oval, the quarter-circumference may be drawn by either of the methods discussed in the First Course.

THE EGG-SHAPED OVAL

In accordance with hygienic considerations, many large storage tanks and containers for industrial products are made with rounded corners instead of sharp, square bends or joints. Fig. 139 shows an outlet connection at the bottom of such a tank on a rounded corner of something like 6 inches radius. The connection is carefully shaped and welded into position. The seams are then hammered and blended smoothly into the tank. The plan of the connection forms an egg-shaped oval, and as this type of oval, so far, has not been dealt with, a few observations on its construction may be helpful.

Egg-shaped ovals may be drawn in a variety of ways. They are generally composed of half a circle and half an ellipse. There is, of course, only one construction for the semicircle, apart from the choice of instruments used, but there are many different methods

of drawing ovals or ellipses, hence the variety of constructions for
egg-shaped ovals.

The example shown in Fig. 135 is a fairly simple method. The
full circle $ABCD$ is first drawn, and then the two straight lines BCF
and DCE. Next, with centre B and radius BD, the arc DF is described. Similarly, with centre D and radius DB, the arc BE is drawn,
and the point C may then serve as the centre for drawing the nose

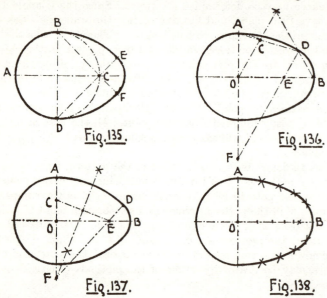

Fig. 135.   Fig. 136.

Fig. 137.   Fig. 138.

arc EF. The chief drawback to this method is that, since the diameters of the oval always bear the same proportion to each other, it
cannot be used with major and minor diameters of any other ratio.

Where the diameters are given, any of the methods of drawing
an ordinary ellipse, either true or approximate, may be used for the
one half of the egg-shaped oval. The illustration at Fig. 136 shows
the half-oval drawn by the method, already dealt with, of describing
an equilateral triangle on the major axis OB, swinging an arc AC
from the centre O, drawing a straight line AD through C, and making
the line DF parallel to CO. Then F is the centre for the arc AD, and
E is the centre for the arc DB.

Perhaps the simplest and most adaptable method of drawing an
egg-shaped oval is that shown in Fig. 137. It is sometimes required
to make the nose radius at the smaller end of the oval either large

or small to suit given conditions. Therefore, after drawing the semi-circle at the large end, describe the nose arc with the required radius *EB* at the other. Next, mark off this radius on the minor axis from *A* to *C*, thus making *AC* equal to *EB*. Join *CE*, and bisect it, producing the bisector to meet the minor axis produced in *F*. Then the point *F* is the centre for the arc *AD*, and *E* is the centre for the arc *DB*. The advantages of this method are that it can be used for diameters of any given dimensions; the nose radius can be made to suit requirements; the construction is simple and easy to remember. On the other hand, the method shown in Fig. 136 gives an oval curve closely approximating to the true ellipse, as will be seen by comparison with Fig. 138.

OUTLET CONNECTION TO TANK

In the plan of the outlet, Fig. 139, the joint line between the horizontal pipe and the oval connecting piece is represented as a circle. This is because the joint line occurs at 45 degrees in the elevation. It is, however, important to note that its true shape is an ellipse with its major axis from 2′ to 14′, and its minor axis equal to the diameter of the circle. The connecting piece is a transformer between the egg-shaped oval top to the elliptical joint line below. The oval top may be drawn by the method of Fig. 137, and since the plan is symmetrical about the horizontal centreline, only one half will be triangulated for the pattern.

First, the true shape of the ellipse should be obtained on 2′,14′ in the elevation. To do this, divide the semicircle in the plan into six equal parts, as from 2 to 14, and project these points vertically upwards to the joint line 2′,14′ in the elevation. From the points obtained on the joint line, draw lines at right angles to it, and cut them off equal in length to the corresponding widths of the semicircle in the plan. Points will thus be afforded through which to draw the semi-ellipse.

Next, divide the half egg-shaped oval in the plan into six parts. In this case divide the quarter ellipse on the one side into three equal parts, as from 1 to 7, and then the remaining quadrant into three equal parts, from 7 to 13. Project all these points vertically upwards to the top edge in the elevation and number them from 1′ to 13′ as illustrated. Now triangulate the surface in the plan and elevation by joining the consecutive points between the top and the bottom by the zigzag line 1,2,3,4, . . . 11,12,13,14. In the elevation erect a vertical height line and on it project all the points on the elliptical joint line from 2′ to 14′, and also those on the top edge from 1′ to 13′.

To develop the pattern, take the distance 1′,2′ from the elevation, and, as this is the true length between those two points, mark off

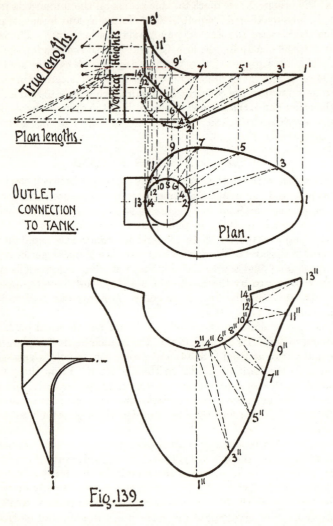

True lengths.

Vertical Heights

Plan lengths.

OUTLET
CONNECTION
TO TANK.

Plan.

Fig.139.

the distance 1″,2″ in the pattern. Next take the plan length 2,3, and mark it off at right angles to the vertical height along the line level with point 2′. Take the true length diagonal up to the point level with 3′, and from 2″ in the pattern, swing an arc through point

3″. Now take the true distance 1,3 direct from the plan, and from 1″ in the pattern describe an arc cutting the previous arc in 3″.

For the second triangle, take the plan length 3,4 and mark it off at right angles to the vertical height along the line level with 4′. Take the true length diagonal up to the point level with 3′, and from 3″ in the pattern swing an arc through point 4″. Now, the true distance between points 2 and 4 is not obtained from the plan, but from the first spacing round the semi-ellipse in the elevation from point 2′. Take this distance in the compasses, and from point 2″ in the pattern describe an arc cutting the previous arc in point 4″.

For the third triangle, take the plan length 4,5 and triangulate it against the vertical height between 4′ and 5′ in the elevation. Take the true length diagonal and from point 4″ in the pattern swing an arc through point 5″. Take the true distance 3,5 direct from the plan and from point 3″ in the pattern describe an arc cutting the previous arc in point 5″.

For the fourth triangle repeat this process with plan length 5,6, but be careful to mark it off at right angles to the vertical height along the line level with point 6′. The true distance between 4 and 6 should be taken from the corresponding spacing on the semi-ellipse.

For the fifth and sixth triangles, repeat the process with plan lengths 6,7 and 7,8; also, take the true distance 5,7 direct from the plan, and the true distance 6,8 from the corresponding spacing on the semi-ellipse. For the seventh triangle take the plan length 8,9, and mark it off at right angles to the vertical height level with point 8′. Take the true length diagonal, this time up to the point level with 9′, and from point 8″ in the pattern swing an arc through point 9″. The next true distance from 7 to 9 cannot be obtained direct from the plan, since the edge now begins to curve upwards. Therefore, the plan length must be triangulated against the vertical height. Thus, take the distance 7,9 from the plan and mark it off at right angles to the vertical height along the line level with 7′, take the true length diagonal up to 9′, and from 7″ in the pattern describe an arc cutting the previous arc in 9″. From this point the rest of the pattern should be easily followed, as the procedure is similar to the end. Care should be taken, however, to triangulate the plan lengths 9,11 and 11,13 against the appropriate vertical heights in order to obtain the true distances between those points.

The smaller diagram in the bottom left-hand corner represents another practical application of the same problem, a transformer connection to a hood or cover with a top corner radius. The edge is either wired or flanged to suit requirements.

AN IMPORTANT PRINCIPLE

It would be a very simple matter to present a problem such as that in Fig. 143, and prepare detailed instructions so that the pattern could be easily developed, and yet leave important principles unexplained. For example, in the plan of the transformer, Fig. 143, the circle represents the top and the rectangle gives the plan form of

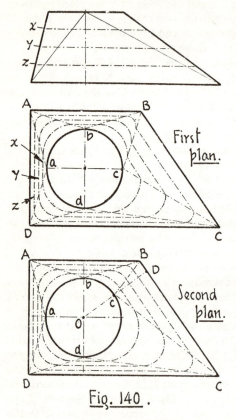

the bottom. In triangulating an ordinary circle-to-rectangle transformer, the surface would be divided up after the manner shown in the smaller figure at (*a*), Fig. 142. That is, each of the four quadrants of the circle would be joined up to the corresponding corner of the rectangle. In the diagram at Fig. 143 it will be seen that each of the four corner points of the rectangle takes only one-sixth of the circle. The remaining two-sixths, on opposite sides, are joined to the middle point of each side of the rectangle, as at point 6, and one might reasonably ask why. This feature embodies an important principle, which can best be explained by reference to Fig. 140.

Fig. 140 .

TRIANGULATING A SMOOTH SURFACE

In triangulating any transforming piece, the object should be to preserve a smooth surface, without introducing kinks or slight bends where they are not wanted. The diagram at Fig. 140 is of an ordinary transformer with one side, *BC*, set at an angle other than 90 degrees. In the elevation, three lines, at *x,y*, and *z*, are represented as drawn horizontally round the surface of the transformer. These lines are again represented in each of the plans as passing round the form of the body. In the first plan it will be seen that the apex of the flat triangular side on BC lies at the quarter point c of the circle. This construction makes the conical corner, Bbc, a full quarter of an oblique cone, similar to the opposite corner, Aab, and the plane of the flat triangle, BCc, forms a kink, or bend, with the conical corner along the line Bc. This kink can be plainly seen by an inspection of the contour lines x,y,z, as also can the reverse kink on the opposite corner Cc.

Now, by moving the apex c of the flat triangle BCc round to a point on the circle which shall lie on the line OD at right angles to BC, the contour lines will present no kinks on Bc and Cc. This will be evident in the second plan, where the contour lines pass smoothly from the conical corners to the flat triangles between them. Hence the correct position of the apex c of the flat triangle is as shown in the second plan.

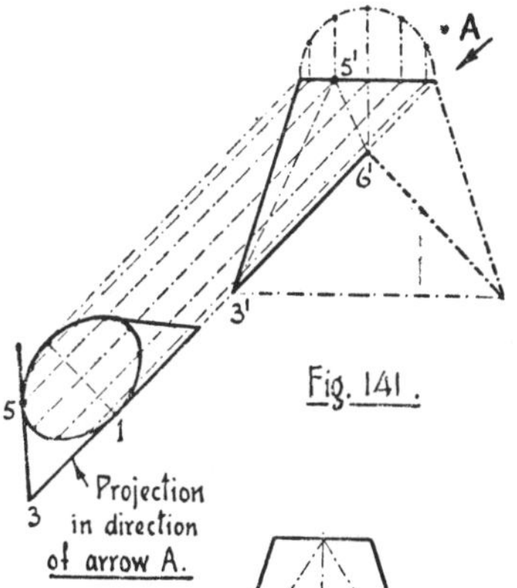

Referring again to Fig. 143, the apex of the triangle 3,5,6 is placed at point 5, and not at the quarter point 7, in order to avoid a constructional kink be-

Fig. 141.

Projection in direction of arrow A.

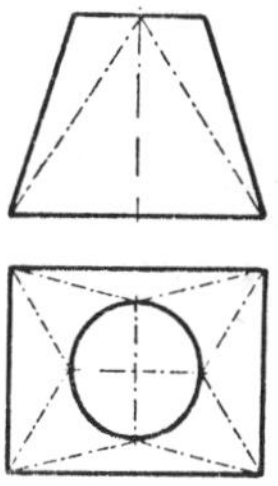

An ordinary transformer.

(a)

Fig. 142.

tween the curved conical corner and the flat triangle. In this particular case the correct location of point 5 is not so easily obtained as point c is in the illustration at Fig. 140. This is because the base 3,6 of the triangle is not in a horizontal plane, but rises from $3'$ to $6'$ as shown in the elevation. The exact position of point 5 may be located as shown in Fig. 141, in which a view is projected in the direction of arrow A, or looking straight down the line $3',6'$. Thus, the circular top is seen as the ellipse, and the straight line $3',6'$ becomes the corner point 3. Half of the width of the transformer is marked off as from 1 to 3

Now, if a plane surface, say a sheet of metal, were placed with its edge along $3',6'$, so that in the projected view it would be seen end-wise as a straight line, it would not be difficult to imagine that sheet to be leaning against the ellipse as represented by the line 3,5, in the projected view. Then the point at which the sheet touches the ellipse is the required point 5. It will be readily seen that the position of point 5 will vary according to the width 1,3 of the transformer, and the relative size and position of the ellipse. In other words, it need not occupy a position such as will divide the top circle into exact sixths, as in this particular example. Nevertheless, this illustration will serve to show the importance of shifting the position of the apex of the flat triangle to suit that of its base.

TRANSFORMER ON ROOF APEX

The type of transformer shown in Fig. 143 is such as may form the base of a ventilator or other outlet on the apex of a roof. Since the plan is symmetrical about the horizontal axis, only one-half of it will be triangulated for the pattern.

First divide the semicircle into six equal parts, from 2 to 11, and join them to points 1,3,6,9,12, as shown in the diagram. Erect a vertical height line in the elevation and project base lines from points 3' and 6'. To develop the pattern, take the plan length 1,2, and mark it off at right angles to the vertical height along the base line level with point 3'. Take the true length diagonal up to the top and mark of 1″,2″ in any convenient position in the pattern. Next take the plan length 2,3, and mark it off at right angles to the vertical height along the base line level with 3'. Take the true length diagonal up to the top, and from point 2″ in the pattern swing an arc through point 3″. Now take the true distance 1,3, direct from the plan, and from point 1″ in the pattern describe an arc cutting the previous arc in point 3″. For the second triangle, take the plan length 3,4 and mark it off at right angles to the vertical height along the base line level with 3'. Take the true length diagonal, and from point 3″

in the pattern swing an arc through point 4″. Next, take the true distance 2,4 direct from the plan and from point 2″ in the pattern describe an arc cutting the previous arc in point 4″. For the third triangle, take the plan length 3,5 and mark it off at right angles to the vertical height along the base line level with point 3′. Take the true length diagonal, and from point 3″ in the pattern swing an arc through point 5″. Next take the true distance 4,5 direct from the plan and from point 4″ in the pattern describe an arc cutting the previous arc in point 5″. For the fourth triangle take the plan length 5,6, and mark it off at right angles to the vertical height, this time along the base line level with point 6′. Take the true length diagonal up to the top, and from point 5″ in the pattern swing an arc through point 6″. Next take the true distance 3′,6′, this time, it is important to note, direct from the elevation and not from the plan, and from point 3″ in the pattern describe an arc cutting the previous arc in point 6″. For the fifth triangle, repeat this process with plan length 6,7, being careful to mark it off against the vertical height along the base line level with 6′. Next, the true distance 5,7 from the plan should complete the triangle.

Since the pattern is symmetrical about the line 6″,7″, the remainder of the pattern to point 12″ is a repetition of this process in the reverse order.

TWO-WAY BREECHES PIECE

One might be excused for jumping to the conclusion that the limbs forming the breeches piece in Fig. 144 were parts of oblique cones. They have the appearance of such, but a little observation on the properties of oblique cones will show that they are not. Assuming that the shapes at AB and CD must be circular, these branch pieces can only be parts of oblique cones when AB and CD are parallel to each other, or when CD occupies the position of the subcontrary section of the oblique cone, which does not often occur, and, in fact, is rarely considered. The subcontrary section of the oblique cone was dealt with in the opening chapter of the Second Course under the properties of the oblique cone. Since the shapes at AB and CD, Fig. 144, are to be considered as circular, the branch piece does not conform to the conditions of the oblique cone, and the pattern cannot be developed by the radial line method. The method of triangulation must therefore be used to obtain the pattern for this branch piece.

Since the plan of the breeches piece, Fig. 144, is symmetrical about the horizontal centreline, only one half will be triangulated for the pattern. The base semicircle is divided into six equal parts, and the points projected vertically upward to the base AB in the elevation. The top edge of the left-hand limb has a semicircle described on it, divided into six equal parts, and the points projected perpendicularly back to the edge 1′,13′. The points on this edge are now joined

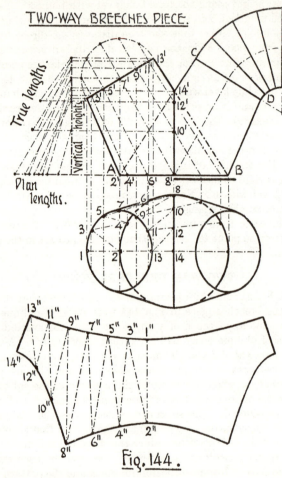

TWO-WAY BREECHES PIECE.

Fig. 144.

to the corresponding points on the base, as represented by the lines 1′,2′; 3′,4′; 5′,6′; 7′,8′; and so on. The top edge 1′,13′ should now be dropped into the plan to obtain the ellipse 1,13. Detailed directions of minor operations such as these are not given in full, as it is assumed that practice on these in connection with previous problems will have made the reader and student familiar with the mode of procedure. The points 1 to 13 on this ellipse in the plan are next joined to the corresponding points on the base circle. Thus, the lines drawn between the top edge and the base are now represented both in the plan and the elevation. So far the left-hand limb has been considered as a complete transformer between the full base AB and the top 1′,13′, but the joint between the two branches cuts this on the line 8′,14′. For this reason, the lines which cross the cut-off must, in triangulating the surface, be considered as terminating on the line 8, 14, as shown at 8,10,12,14 in the plan and at 8′,10′,12′,14′ in the elevation. Beginning at the outside point on the centreline, divide the surface into triangles by inserting diagonals between the lines already drawn, thus forming the zigzag line as shown in the plan at 1,2,3,4, . . . 11,12,13,14. The zigzag line is not shown in the elevation for the sake of avoiding congestion of lines, but the corresponding points are numbered 1′,2′,3′,4′, . . . 11′,12′,13′,14′. Erect a vertical height line in the elevation, and project all the points at the top edge 1′ to 13′, horizontally to it, and also those on the joint line 8′,10′,12′,14′.

To develop the pattern, the first line 1′,2′ may be taken direct from the elevation, since that is its true length, and marked off in any convenient position in the pattern, as at 1″,2″. Next take the plan length 2,3, and mark it off along the base line at right angles to the vertical height. Take the true length diagonal up to the point on the vertical height line level with 3′, and from point 2″ in the pattern swing an arc through point 3″. The next true distance 1″,3″ should not be taken from the plan, but from the semicircle on the top edge 1′,13′. The six equal divisions round the semicircle are the true distances required for the top edge in the pattern. Therefore, take one of those divisions and from 1″ in the pattern describe an arc cutting the previous arc in point 3″.

For the second triangle, take the plan length 3,4 and mark it off at right angles to the vertical height. Take the true length diagonal up to the point level with 3′, and from point 3″ in the pattern swing an arc through point 4″. Next take the true distance 2,4, direct from the plan, and from point 2″ in the pattern describe an arc cutting the previous arc in point 4″.

For the third, fourth, fifth, sixth, and seventh triangles, repeat this process with plan lengths, 4,5; 5,6; 6,7; 7,8; and 8,9, being careful to triangulate them against vertical heights level with points 5,′7′, and 9′ respectively. Also, take the true distances 3″,5″; 5″,7″; 7″,9″, direct from the semicircle in the elevation, and the true distances 4″,6″ and 6″,8″ direct from the corresponding distances in the plan.

Now, for the eighth triangle, take the plan length 9,10, and mark it off at right angles to the vertical height, this time along the base line level with 10′. Take the true length diagonal up to the point level with 9′, and from 9″ in the pattern swing an arc through point 10″. Next, in the pattern, the true distance 8″,10″ must be obtained, and it is important to note this change. Take the plan distance 8,10, which is the first spacing on the joint curve, and mark this off at right angles to the vertical height along the bottom base line. Take the true length diagonal up to the point level with 10′, and from 8″ in the pattern describe an arc cutting the previous arc in 10″.

For the ninth triangle, take the plan length 10,11, and mark it off at right angles to the vertical height along the second base line level with 10′. Take the true length diagonal up to the point level with 11′, and from point 10″ in the pattern swing an arc through 11″. Next, take a true division from the semicircle in the elevation and from point 9″ in the pattern describe an arc cutting the previous arc in 11″.

For the tenth triangle, take the plan length 11,12, and mark it off at right angles to the vertical height along the third base line level with point 12′. Take the true length diagonal up to the point level with 11′, and from point 11″ in the pattern swing an arc through 12″. Next take the plan length 10,12, which is the second spacing on the joint curve, and mark this off at right angles to the vertical height along the second base line level with 10′. Take the true length diagonal up to the point level with 12′, and from point 10″ in the pattern describe an arc cutting the previous arc in point 12″.

The remaining two triangles should be easily plotted by following these directions with plan lengths 12,13 and 13,14, and taking the true distances 9″,11″ and 11″,13″ from the semicircle in the elevation. The final true distance 12″,14″ is again obtained by triangulating the plan length 12,14 against the appropriate vertical height.

FOUR-WAY BREECHES PIECE

The problem of the four-way breeches piece represented in Fig. 145 is similar in general principles to the two-way example just dealt with, but in this case the joint line in the plan cuts across at 45 degrees, which presents a curve in the elevation instead of a straight line. This curve must be determined before the pattern can be developed. In this example, one limb only is considered, and is regarded as a transformer between the full circular base and the top, but cut off at the position of the joint lines, shown in the plan from 6 to 14, and 14 to 15. As in the previous example, the semicircle of the base is divided into six equal parts, $a,b,c,d,e,4,2$, as also is the corresponding half of the top from 1 to 13. The points on the base

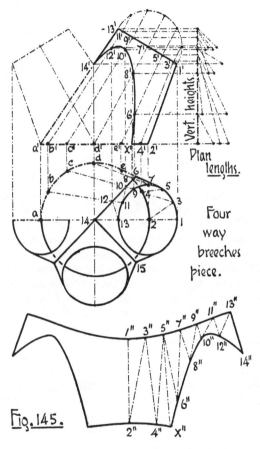

Vert. heights.

Plan lengths.

Four way breeches piece.

Fig. 145.

and the top are then joined as in the diagram at Fig. 145, both in the plan and elevation. Then the points where these lines cross the joint line in the plan, as, for instance where the line $11,b$ crosses the joint line in point 12, should be transferred vertically upwards to the corresponding lines in the elevation. Thus, the points $6',8',10',12',14'$ will be obtained through which to draw the required joint curve.

One other point, X, remains to be located. This, in the plan, falls on the base circle between points 4 and e, and apparently coincides with point 8, which is higher up on the joint line. The correct point is that in which the base circle crosses the joint line, and this transferred vertically upwards to the base line in the elevation will give point X'.

The method of triangulating the surface of the branch piece is similar to that of the previous example, and should be readily followed from the diagram. The development of the pattern also is similar and the description would merely constitute a repetition of the directions given for the two-way breeches piece, except that the joint curve in the pattern begins at point X'' between $4''$ and $6''$, and an extra line, $5'',X''$, needs to be triangulated between $4'',5''$ and $5'',6''$.

THE JUNCTION PIECE

The particular type of branch piece shown in Fig. 146 should appeal to craftsmen who are interested in efficiency, not only from the constructive point of view, but also from that of easy air flow. The design of a branch piece is often governed by the craftsman's ability to visualize the requirements, and also depends a good deal on his skill in pattern-drafting. This design, in order to differentiate from the breeches piece dealt with in the previous section, may be called the "junction piece." It does not contain a joint in the middle between the branches other than the short one from $11'$ to $12'$ in the elevation. The pattern is bent along the lines $8',11'$ and $8',12'$, which thus avoids the awkward curved joint which occurs in the corresponding position between the limbs of the breeches piece. This type of junction piece admits of a wide range of variation to suit circumstances. Not only two-way junctions, but three-way, four-way, and even six-way pieces can be made, which may transform from a circular base to the circular branches at the other end, or from a square or rectangular base to circular branches. The problem given in Fig. 146 is of a two-way junction piece which branches equally both ways, while that of Fig. 147 is one in which one branch continues in the same direction as the main duct and the other takes out from the side.

TWO-WAY JUNCTION PIECE

The plan of the junction piece shown in Fig. 146 is symmetrical about the horizontal centreline and also about the vertical centreline. On this account only one quarter of the plan is triangulated for the

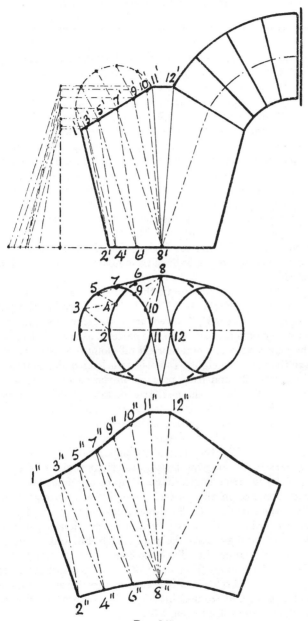

FIG. 146

pattern. It is assumed that the setting out of the plan and elevation is understood.

To divide the surface into triangles, describe a semicircle on the top edge 1′,11′, of one branch in the elevation, divide it into six equal parts and project the points perpendicularly back to the edge. Drop these points into the plan to obtain the divisions on the semi-ellipse, and number them 1,3,5,7,9,10,11, as shown in the diagram; also number the opposite corner of the middle triangle, 12. Next divide the quarter-circle of the base into three equal parts and number the points 2,4,6,8. Now triangulate the surface by joining up the points as shown in the plan and elevation. It will be seen that the points on the quarter-circle of the base are joined to the corresponding points on the quarter-circle at the top, and from point 8 on the base a number of lines radiate to the other quarter-circle at the top, thus forming a quarter-cone.

This method of dividing the surface into triangles should be carefully noted, as it forms the basis of triangulating all kinds of junction piece of this class.

Erect a vertical height line in the elevation, and project all the points on the top edge 1′ to 11′ on to it. To develop the pattern, take the true distance 1′,2′, direct from the elevation and mark it off in any convenient position in the pattern, as from 1″ to 2″. Next take the plan length 2,3, and mark it off along the base line at right angles to the vertical height. Take the true length diagonal up to the point level with 3′, and from point 2″ in the pattern swing an arc through point 3″. Now, the next true distance 1″,3″ in the pattern is obtained from the semicircle on the top edge 1′,11′ in the elevation. Since the divisions on the semicircle are equal and represent true distances, one of these should be marked off in the pattern from 1″ to 3″. For the second triangle, take the plan length 3,4 and mark it off along the base line at right angles to the vertical height. Take the true length diagonal up to the point level with 3′, and from point 3″ in the pattern swing an arc through point 4″. Next, take the true distance 2,4 direct from the plan, and from point 2″ in the pattern describe an arc cutting the previous arc in point 4″. For the third triangle take the plan length 4,5, and triangulate it against the vertical height. Take the true length diagonal up to the point level with 5′, and from point 4″ in the pattern swing an arc through point 5″. Next take a true spacing direct from the semicircle in the elevation, and from point 3″ in the pattern describe an arc cutting the previous arc in point 5″.

For the rest of the pattern this process should be repeated with

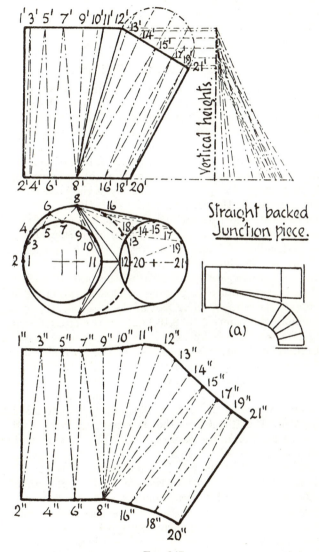

Straight backed
Junction piece.

(a)

Vertical heights

Fig. 147

plan lengths 5,6; 6,7; 7,8; 8,9; 8,10; 8,11; 8,12, and for the true distances 5″,7″; 7″,9″; 9″,10″; 10″,11″, round the top in the pattern, the spacings from the semicircle in the elevation should be taken. For the remaining true distances round the bottom, the divisions 4,6 and 6,8 should be taken direct from the plan.

STRAIGHT-BACKED JUNCTION

The particular feature of the junction piece shown at (*a*), Fig. 147, is that its back maintains a straight line with the direction of the main duct. This has advantages in relation to air flow, and also favourable points in construction. It also has a pleasing appearance when made up, and, while this point may not be regarded as of economical value, it is, nevertheless, a point well worth consideration. However, it is a minor point compared to its other advantages.

The plan in this case is symmetrical about the horizontal centre-line, but not about the vertical. One half is therefore triangulated for the pattern. Describe a semicircle on the top edge of the branch which is set at an angle, divide it into six equal parts and project the points perpendicularly back to the edge. Drop these points into the plan to obtain the ellipse, and also the divisions on it from 12 to 21. Divide the corresponding semicircle of the other branch into six equal parts and number the points 1 to 11 as shown in the diagram. Also divide the corresponding half of the base into six equal parts and number the points 2,4,6,8,16,18,20, in the order shown. Now triangulate the surface by joining up the points as shown in the diagram. The elevation is also correspondingly divided up to help in making the method more clear. It is similar to that of the previous problem, although it may appear a little more complex.

Erect a vertical height line in the elevation and project all the points on the edge 12′,21′ on to it. To develop the pattern, take the true distance 1′,2′ direct from the elevation, and mark it off in any convenient position in the pattern, as at 1″,2″. Next, take the plan length 2,3, and mark it off along the base line at right angles to the vertical height. Take the true length diagonal up to the top point, and from point 2″ in the pattern swing an arc through point 3″. Now take the true spacing 1,3 direct from the plan, and from point 1″ in the pattern describe an arc cutting the previous arc in point 3″. For the second triangle, take the plan length 3,4, and mark it off along the base line at right angles to the vertical height. Take the true length diagonal up to the top point, and from point 3″ in the pattern swing an arc through point 4″. Next, take the true distance 2,4, direct from the base semicircle in the plan, and from point 2″ in the pattern describe an arc cutting the previous arc in point 4″. For the third triangle, repeat this process with plan length 4,5, and also with the true distance 3,5 direct from the plan.

There are nineteen triangles in all, but if this process is followed carefully, no difficulty should be experienced in completing the

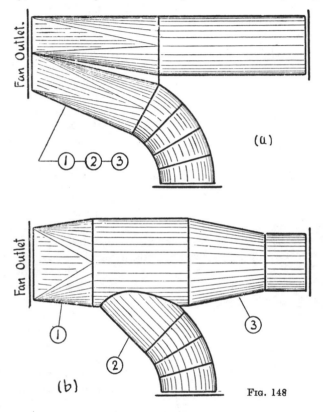

(a)

(b)   FIG. 148

pattern, since the process is the same right through. Care should be exercised, however, in the second half of the pattern from 12″ to 21″, to take the true lengths against the vertical height up to the corresponding points level with those on the edge 12′,21′.

THREE UNITS IN ONE

Geometry as a source of economy or efficiency in sheet metal work is a very effective means of saving time. Efficiency is execution with ease and simplicity; a branch piece, easy to make, which transforms from a square or rectangle to two or three circles, and adjusts the area to suit, all in one unit, is an example of this.

The branch piece shown in Fig. 148 (a) embodies three units in one, and transforms from a square at the base to two circles at the other end. The common method of achieving the same object is to make a square-to-circle transformer, take out the branch in the form of an oblique stump, and then adjust the area by inserting a

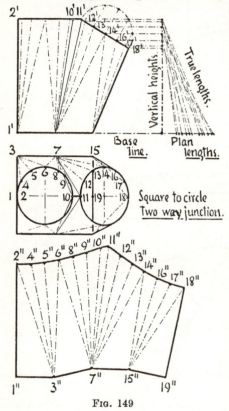

reducing cone. This is illustrated in Fig. 148 (b). There is less work in the transformer branch piece at (a) than in the three units at (b), and the branch piece is more in keeping with the theoretical principles of air flow.

The example given in Fig. 150 represents a three-way junction on the same principles, transforming from a rectangular base to three circular holes above. The middle hole, larger than the others, forms the outlet for the continuation of the main duct, while the two side holes each take a branch pipe. This three-way junction is made in two pieces of metal. It can be made in one piece only if desired, but it is easier to make it in two halves.

FIG. 149

SQUARE-TO-CIRCLE TWO-WAY JUNCTION

The problem shown in Fig. 149 is of a straight-backed junction, and is not unlike a double tallboy transformer, but there is no joint between the limbs other than a short seam at the top of the middle triangles, as from 10' to 11'. The pattern is bent along the lines 7″,10″ and 7″,11″, and then shaped and formed so that the seams come in the positions of the lines 1″,2″ and 18″,19″.

Preparatory to developing the pattern, describe the semicircle on the top edge of the limb, which is inclined at 30 degrees, divide it into six equal parts and project the points perpendicularly back to the edge. Drop the points on the edge vertically into the plan to obtain the corresponding points 11 to 18, on the ellipse. Now divide the plan into triangles as shown in the diagram, and number the points from 1 to 19. Erect a vertical height line in the elevation, and project all the points, 11' to 18', on the inclined edge horizontally on to it.

To develop the pattern, take the vertical depth 1′,2′, direct from the elevation, and mark it off in any convenient position in the pattern, as at 1″,2″. Next take the plan length 2,3, and mark it off along the base line at right angles to the vertical height. Take the true length diagonal up to the top, and from point 2″ in the pattern swing an arc through point 3″. Now take the true distance 1,3 direct from the plan, and from point 1″ in the pattern describe an arc cutting the previous arc in point 3″. For the second triangle, take the plan length 3,4, and mark it off along the base line at right angles to the vertical height. Take the true length diagonal up to the top, and from point 3″ in the pattern swing an arc through point 4″.

Next take the true distance 2,4, direct from the plan and from point 2″ in the pattern describe an arc cutting the previous arc in point 4″. For the third triangle, take the plan length 3,5, and mark it off along the base line at right angles to the vertical height. Take the true length diagonal up to the top, and from point 3″ in the pattern swing an arc through point 5″. Next take the true distance 4,5, direct from the plan and from point 4″ in the pattern describe an arc cutting the previous arc in point 5″.

The remainder of the pattern should not be difficult to develop from this point if the method is carefully followed. The true spacings along the top of the pattern from 2″ to 11″, are taken direct from the plan, but the remaining true distances from 11″ to 18″ are obtained from the semicircle on the top edge of that limb in the elevation. Also, the true distances along the bottom of the pattern, from 1″ to 19″, are taken direct from the plan. One other word of caution: the plan lengths in the second part of the problem must be triangulated against the appropriate vertical heights from the top edge 11',18' in the elevation.

THREE-WAY JUNCTION, RECTANGLE TO CIRCLE

The problem of the three-way junction piece, transforming from a rectangle to three circles, as shown in Fig. 150, illustrates how this method may be extended, even to junctions of four or more branches. The pattern-development is much the same as for the two-way junction, but in this case the plan is symmetrical about the horizontal and also the vertical centre lines, and therefore only one quarter is triangulated for the pattern.

As in the previous problem, describe a semicircle on the top edge of the limb, which is inclined at an angle, divide it into six equal parts and project the points perpendicularly back to the edge 2′,10′. Drop the points on the edge vertically into the plan to obtain the corresponding points, 2 to 10, on the ellipse. The quarter-circle 11,14 should now be divided into three equal parts, and the surface of the junction divided into triangles by numbering the points from 1 to 15 as shown in the diagram. Erect a vertical height line in the elevation, and project all the points on the edge 2′,10′ on to it.

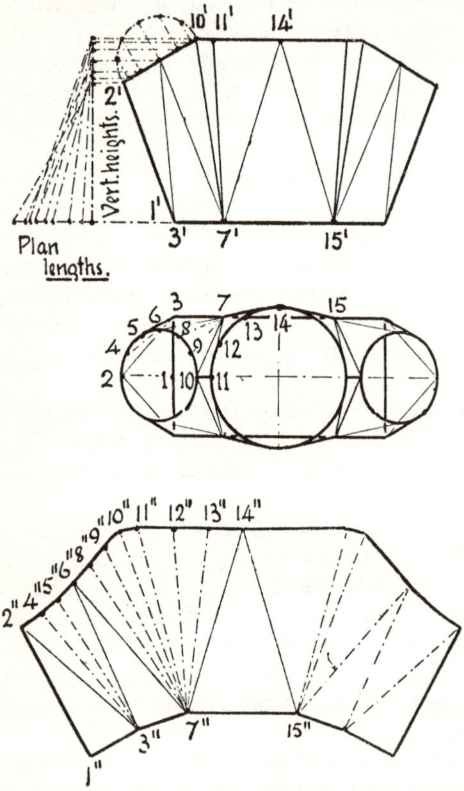

Fig. 150

To develop the pattern, take the true distance 1′,2′ direct from the elevation and mark it off in any convenient position in the pattern, as at 1″,2″. Next take the plan length 2,3, and mark it off along the base line at right angles to the vertical height. Take the true length diagonal up to the bottom point level with 2′, and from point 2″ in the pattern swing an arc through point 3″.

Now take the true distance 1,3 direct from the plan, and from point 1″ in the pattern describe an arc cutting the previous arc in point 3″. For the second triangle, take the plan length 3,4, and mark it off along the base line at right angles to the vertical height. Take the true length diagonal, this time up to the point level with point 4′, or the second point up, and from point 3″ in the pattern swing an arc through point 4.″ The next true distance, 2″,4″, should not be taken from the plan but from the semicircle on 2′,10′ in the elevation. Take one of those spacings and from point 2″ in the pattern describe an arc cutting the previous arc in point 4″. For the remainder of the pattern, follow this method carefully, making sure of each true length against the appropriate vertical height.

There is one point, however, which may give rise to query, and that is the position of point 7 on the side of the rectangular base. Unless special considerations determine otherwise, the combined areas of the circles at the top should be equal to the area of the rectangle at the base. The distance 3,7 should be such that the area enclosed by that and the width across the base is equal to the area of the circular branch which it subtends. Thus, the full rectangular base is, in effect, divided into three smaller rectangles, of areas corresponding to those of the circular branches above. The position of point 7 is therefore determined by the relative areas of the branches.

FOURTH COURSE

CHAPTER 11

METHOD OF CUTTING PLANES

METHODS of pattern-drafting have, up to now, received a good deal of attention and, it is hoped, elucidation. In this Fourth Course of geometry applied to sheet metal work, the solution of problems of

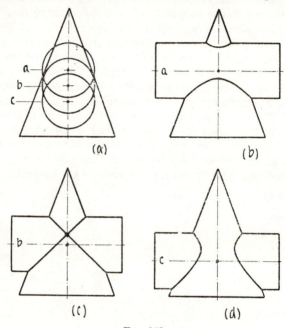

(a)

(b)

(c)

(d)

FIG. 151

intersection and interpenetration will be dealt with on systematic lines, emphasizing the importance of similarity of method.

The determination of the joint line between two intersecting bodies is not always an easy problem, and in practice accuracy is often dismissed in favour of a guess. While approximation may sometimes pass with credit where the shape of an edge is not important, it does not generally pay when joint lines of mathematical precision have to be dealt with. Take, for example, the line of intersection between a cone and a cylinder, as represented in the diagram at Fig. 151. The joint line between a given pair of intersecting bodies forms a definite contour which can be determined accurately. The shape of the curve is an exact line from which deviation cannot be made without error. Therefore, when a pattern for an intersecting body is drafted by guesswork either it must be carefully trimmed so that the joint curve ultimately reaches its correct form, or the intersecting bodies must be distorted to allow the faulty joint lines to meet. Trimming is too easily overdone and is apt to finish on the wrong side of the line.

VARYING LINE OF INTERSECTION

Although a cone and cylinder intersecting under given conditions produce a definite joint line, the form of that line will vary according to the relative positions of the central axes. The importance of this will be seen by reference to Fig. 151. The side view in the top left-hand position (a) shows a cone and a cylinder as seen endwise. The cylinder is shown with its central axis in three different positions. In the bottom position the cylinder penetrates the cone. In the middle position the cylinder and cone intersect tangentially, that is, the circle representing the cylinder just touches the sides of the cone. In the top position the cone penetrates the cylinder. The other diagrams at (b), (c), and (d) show front views of the three different positions and illustrate the effect on the form of the joint line.

CLASSIFICATION OF METHODS

Before the pattern can be developed the line of intersection must be determined. In much the same way that methods of development can be classified under the Radial Line method, the Parallel Line method, and Triangulation, so the processes of determining lines of intersection can be grouped under the Method of Cutting Planes, the Method of Projection, and the Method of Concentric Spheres. The principles on which these are based may be used in solving any problem of intersection. When examples are treated as independent problems, the processes are often trimmed so that they bear no resemblance to one another nor to any general method. It is, however, better to follow a systematic study, in which similarities are emphasized even though it involves a few more lines and a little extra work. The reward is a clearer understanding and greater facility in application. A number of problems of intersection have already been dealt with, such as pipe tee pieces and intersection around a common central sphere. The latter, however, should not be confused with the Method of Concentric Spheres mentioned above.

METHOD OF CUTTING PLANES

When a regular geometrical body is cut by a plane, a definite form is presented at the cutting plane. For example, when a cone is cut by a plane parallel to its slant, as at *OB*, Fig. 152, the form of the cut, or the section at the cutting plane, is a parabola. When the cutting

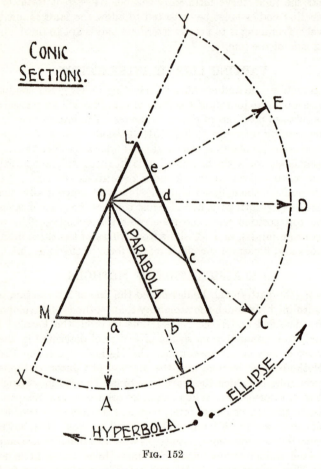

Fig. 152

plane occurs in any position in clockwise rotation from point *O*, the form of the cross-section is a hyperbola. When the cutting plane occupies any position in anti-clockwise rotation from point *O*, the form of the cross-section is an ellipse. It will be observed, however, that when the cutting plane is at right angles to the central axis of the cone, as at OD, the form of the cut is a circle. This occurs among the ellipses, and the circle may be considered as an ellipse with equal axes.

In the case of a cylinder, any cutting plane parallel to its central axis will present a rectangular section, but a cutting plane inclined at an angle, or cutting across the cylinder will present an ellipse. Here again, the cross-section at right angles to the central axis is a circle, and may be considered as a special form of ellipse.

The sphere presents a circle when cut through in any position or in any direction by a plane.

Suppose now that a cylinder penetrating a cone be severed by a cutting plane which passes through both bodies, as illustrated in Fig. 153. The circular edge of the conic section will meet or cross the rectangular edge of the cylinder in four points, as at a,b,c,d. This fact forms the basis of the method of cutting planes in solving

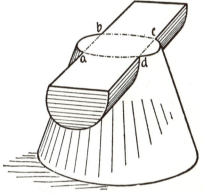

FIG. 153

problems of intersections. The application of this principle is illustrated in Fig. 154, in which front and side elevations and a plan are given of the interpenetration of a cone and cylinder. In this example it is intended to show that a point on the line of intersection may be correctly located no matter in what direction through that point a cutting plane be taken. In solving a problem, however, the cutting planes will be taken in the most convenient positions for easy solution.

In the front elevation, Fig. 154, three cutting planes are shown at AA, BB, and CC. Taking first the horizontal plane through AA, which passes through one of the division points on the semicircle at the end of the cylinder, it will readily be seen that the plan of the cylinder at the position of the cutting plane will be a rectangle, as shown by the chain dotted lines at $A'A'$, and the plan of the cone at the same cutting plane will be a circle as represented at $a'a'$. The circle cuts the rectangle in four points, one of which is marked X. If these four points are now projected vertically upwards to the cutting plane AA, in the elevation, they will give two points thereon, each of which lies on a curve of intersection.

For the purpose of following up this explanation, only one of those

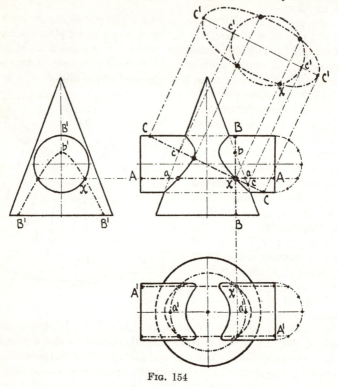

Fig. 154

points, that marked X, will now be considered. Suppose that, instead of the horizontal cutting plane AA, the vertical cutting plane BB were taken. The "full face" view of this cutting plane must be projected into the side elevation, and here the form presented by the cone is the hyperbolic curve $B'b'B'$, and the cylinder presents the full circle. If now the points in which the hyperbola crosses the circle be projected back to the cutting plane BB in the front elevation the same point X will be located.

Again, suppose that yet another cutting plane, CC, be taken through X, inclined at any angle. The "full face" view of the cross-sections of the cone and cylinder is projected at right angles to the plane CC, and the form presented by the cylinder is that of the longer ellipse $C'C'$, while the shape of the cone is represented by the shorter ellipse $c'c'$. It will be seen from the diagram that the points in which the two ellipses cross, when projected perpendicularly back to the cutting plane CC, give points on the intersection curves. Thus the point X in the elevation could again be located from this projection.

It can be shown that point X on the intersection curve in the elevation could be located by a projection on any other cutting plane passing through that point. The important condition to remember is that the true form of the cone and cylinder at the position of the cutting plane must be represented in the projection. Then the points of the crossing must be projected perpendicularly back to the cutting plane to locate the point on the joint curve.

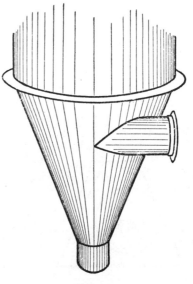

FIG. 155

CONICAL BASE OF AN AIR FILTER

Of the three methods by means of which all problems of intersection may be solved, that of cutting planes is perhaps the most adaptable. In applying this method to the determination of the joint line between a cone and a cylinder, a series of planes, not necessarily horizontal nor parallel to each other, are supposed to cut through both bodies. At each cutting plane two or more points may be located which lie on the line of intersection. The principles underlying this method have already been explained in connection with Fig. 154, and may now be applied in solving such problems as that represented in Fig. 155, which is the conical base of an air filter. The inlet pipe enters the cone tangentially, and also horizontally or parallel to the upper base of the cone.

CONE AND CYLINDER, TANGENTIAL INTERSECTION

The problem given in Fig. 156 is similar to that of the air filter base, except that it is downside up and the inlet pipe is relatively larger. These differences do not alter the problem, but assist in making the processes of solution clearer to follow. It will be observed in Fig. 156 that the patterns are drafted from the front elevation. The

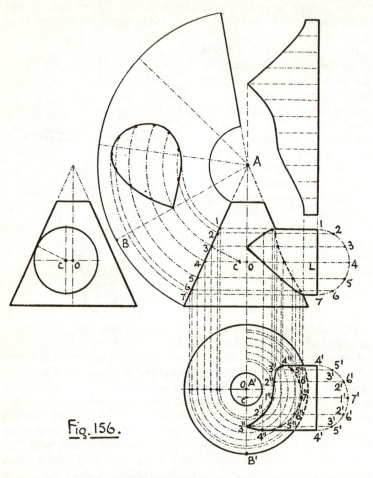

Fig. 156.

correct form of the joint line in that view must therefore be determined prior to developing the patterns. The position of the cylinder relative to the cone is determined from the side elevation. The diameter of the cylinder and the height from the base of the cone are usually given. The horizontal position *OC* must then depend on the tangential relation of the cone to the cylinder. This condition cannot be obtained from the front elevation nor from the plan, and as it is essential to locate in the plan the correct off-centre position OC of the cylinder relative to the central axis of the cone, the side elevation is important.

Nevertheless, the full side elevation may be dispensed with and the necessary off-centre distance obtained from the front elevation by treating the left-hand half of the cone as a side view. Thus, the point C may be obtained in the front elevation by marking off the radius of the cylinder at right angles to the side of the cone so that it cuts the centreline CL in point C. Then CO is the off-centre distance required, which should be used in the plan to mark off the distance OC for the centreline of the cylinder.

Having now determined the position of the cylinder in the elevation and plan, describe semicircles on the end of the cylinder in both views. It must be remembered that these semicircles really represent projected views of the circular end of the cylinder. Therefore, in numbering the points from 1 to 7 in the elevation, as shown in the diagram, it will be seen that those same points in the plan must bear the numbers in the positions shown. That is, the top point 1 in the elevation occupies the middle position in the plan, as at 1', and the other points, proceeding outwards both ways to point 4' at each end, double back over the same points to reach the middle again in point 7'.

From each of the points 1 to 7 in the elevation, draw horizontal lines right through to the other side of the cone. Let these lines represent cutting planes through both bodies. The plan of the cone at each of these cutting planes will be a circle, and the diameters may be obtained by dropping lines from points 1,2,3,4,5,6,7 on the side of the cone vertically downwards to the horizontal centreline in the plan. From the centre O of the cone in the plan, circles may be drawn from these points to represent the cutting planes. In the diagram at Fig. 156 three-quarter circles only are drawn, which is sufficient for this problem. The next step in the plan is to draw lines on the cylinder parallel to its axis from the points on the semicircle at the end. These lines determine the plan of the cylinder at the various cutting planes. For example, consider the cutting plane 3,3; at this cutting plane in the plan the two lines drawn from points 3' and 3' represent the plan width of the cylinder at that plane. Similarly, the two lines drawn from points 6' and 6' represent the plan width of the cylinder at cutting plane 6,6, and so on.

The points of intersection in the plan may now be located. Starting in the elevation at cutting plane number 1 on the side of the cone, follow the line down into the plan and round the circle to the point where the circle intersects the line drawn from point 1'. Mark that point of intersection as number 1″. Now repeat this process by following the line from point 2 in the elevation down into the plan and round the circle to the points where the circle intersects the two lines drawn from points 2' and 2'. Mark these points 2″ and 2″. This process should be repeated from the remaining points 3,4,5,6,7 in the elevation. A line now drawn through the points thus obtained in the plan will give the form of the joint line. With a little practice in locating these points, the process of following the lines down from the elevation will become unnecessary, as they may be marked off easily in the plan at sight.

The line of intersection in the elevation may now be obtained from that in the plan. The points on the intersection line in the plan should be projected vertically upwards to the corresponding cutting planes in the elevation. Thus, points 2″ and 2″ in the plan should be projected upwards vertically to the cutting plane 2,2, and points 3″ and 3″ in the plan should be similarly projected to cutting plane 3,3. Each of the remaining points on the intersection line in the plan should be treated likewise, thereby obtaining sufficient points in the elevation to draw in the form of the line of intersection.

The pattern for the cylindrical part may now be "unrolled" in accordance with the parallel line method of development. The points on the line of intersection should be projected into the pattern to plot the contour of the joint line as shown in the diagram. The pattern for the conical part should present no difficulty, except, perhaps, in the shape of the hole which forms the joint line with the cylinder. To determine the contour of this hole, describe an arc from each of the points 1,2,3,4,5,6,7 on the side of the cone, using the apex A as centre. These arcs represent the form and positions of the cutting planes in the pattern. Draw the radial line AB from the apex A through the arcs to the base line of the cone. Let this line represent, in the pattern, the position of the line $A'B'$ from the plan. Since, from the line $A'B'$ in the plan, the distances *round the curves* to the points 1″,2″,3″,4″,5″,6″,7″ are true lengths, these true distances should be marked off from AB round the corresponding curves in the pattern. A line drawn through these points will give the contour of the hole.

When the principles involved in this method of solving problems of intersection are properly understood, much of the constructional detail may be omitted, but abbreviation is not advisable until such principles can be applied with confidence.

A SHORT-CUT METHOD

The method of solving the problem given in Fig. 157 is based on that of cutting planes, but the arrangement of the constructional details considerably shortens the work in the solution. This problem is similar to that of Fig. 156, except that the cylinder is on-centre with the cone, and the solution is therefore somewhat more simple.

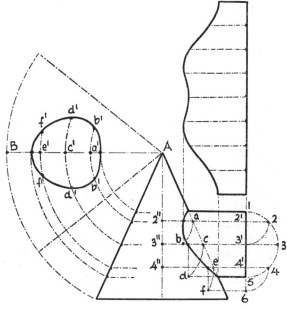

FIG. 157

The semicircle on the end of the cylinder is divided into four equal parts instead of six, and numbered 1,2,3,4,5. Lines from points 2,3, and 4 are projected horizontally back to the centreline of the cone. The resulting points 2″,3″,4″ on that centreline are used respectively as centres for describing the arcs *ab*, *cd*, *ef*. Also, points 2′,3′,4′ are used respectively as centres for describing the quadrants 2,3′; 3,5; 4,6. The arc *ab* cuts the horizontal line produced from the quadrant 2,3′ in point *b*. Project point *b* vertically upwards to line 2,2″, which thus locates a first point on the intersection line. The arc *cd* cuts the horizontal line produced from the quadrant 3,5 in point *d*. Project point *d* vertically upwards to the line 3,3″, which thus locates a second point on the intersection line. The arc *ef* cuts the horizontal

line produced from the quadrant 4,6 in point f. Project point
vertically upwards to the line 4,4″, which thus locates a third point
on the intersection line. A curve now drawn through these points,
as shown in the diagram, will give the form of the joint line. The
pattern for the cylindrical part may now be "unrolled" in the usual
way. To plot the contour of the hole, describe arcs from the
corresponding points on the opposite side of the cone, as shown
in the diagram. Draw a centreline AB to cross the series of arcs,
and measure off along the curves on either side of the points
$a′,c′,e′$, distances equal in length to the arcs ab, cd, ef. A curve
drawn through the points thus obtained will give the contour of
the hole.

That this solution is based on the method of cutting planes may be
seen from the following observations. The curves ab, cd, and ef
constitute portions of the plan of the cone at the positions of the
cutting planes 2,2″; 3,3″; and 4,4″. The perpendicular distances of
points b,d, and f from the respective cutting planes constitute the
semi-width of the plan of the cylinder at those cutting planes. Therefore, the vertical projection of those points back to the corresponding
planes locates points on the line of intersection. This short-cut
method, as given in Fig. 157, applies only to the right cone intersected by a cylinder which has its centreline passing through that of
the cone. It cannot be applied in any simplified form to the tangential problem of Fig. 156, and should be regarded, like all short-cut
methods, as a special application.

METHODS OF GENERAL APPLICATION

However unlikely it may seem that some problems of intersection
find application in sheet metal work, the unexpected sometimes
happens. When this occurs undue delay or difficulty may be avoided
if it can be seen that the solution depends on principles common to
many other examples. The study of a method which has a general
application is more profitable than any attempt to memorize the
solutions of individual problems, and, with this in view, particular
attention is given to the method applied in working the following
solutions in preference to simplification by short cuts. Most attempts
to simplify the solution of a problem result in either an incorrect
method or an abbreviation which makes the solution applicable only
to that problem. The former is inexcusable and the latter involves
undue tax on the memory.

The problems of intersection of the right cone and sphere, and
the oblique cone and sphere, given in Figs. 159 and 160, are solved

by the method of cutting planes. It is important to observe, in the example of the right cone and sphere, that the central axis of the cone does not pass through the centre of the sphere. When the condition occurs that the axis of the cone does pass through the centre of the sphere, the intersection line is a circle whose plane is at right angles to the central axis of the cone; thus, the cone is complete and no problem of intersection arises.

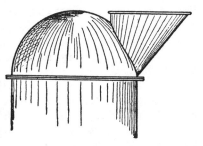

Fig. 158

The illustration at Fig. 158 is a pictorial example of the problem given in Fig. 160.

CONE AND SPHERE INTERSECTION

The first part of the problem presented in Fig. 159, before the pattern can be developed, is to determine the

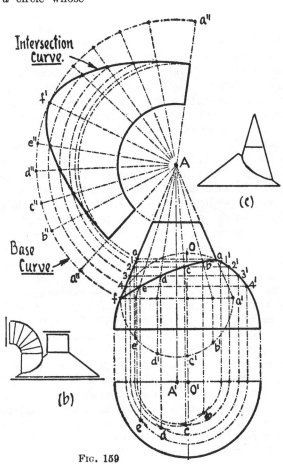

Fig. 159

line of intersection in the elevation. The cone f,A,a' is shown penetrating the dome, the extreme points of contact occurring at a and f. Between the point a and f, four horizontal cutting planes, $1,1'$; $2,2'$; $3,3'$; $4,4'$, are drawn in any positions, cutting through both bodies. The plan of the cone and the plan of the sphere are both circles at each of these cutting planes. In the diagram at Fig. 159, a half-plan of the sphere is given, and the plans of the cutting planes are obtained by dropping vertical lines from the points $1',2',3',4'$ to the horizontal centreline of the sphere in the plan, and describing arcs or semicircles, using point O' as centre. The corresponding plans of the cone and the cutting planes are obtained by dropping vertical lines from points $1,2,3,4$ to the horizontal centreline of the sphere in the plan, and describing arcs or semicircles, this time using point A' as centre. These arcs intersect the corresponding arcs from points $1',2',3',4'$ in points b,c,d,e in the plan. From these points of intersection, lines are projected vertically upwards to the respective cutting planes in the elevation, thus obtaining points b,c,d,e, which occupy positions on the joint curve, or line of penetration. A curve now drawn from point a, through b,c,d,e to f, gives the required form of the line of intersection.

To develop the pattern for the cone, describe a semicircle on the base f,a'. Next, from the apex A, draw lines on the cone through points b,c,d,e to the base f,a'. From the points obtained on the base draw perpendicular lines to cut the semicircle in points b',c',d',e'. These points on the semicircle represent the plan positions of the points on the base line f,a'. Now, from the apex A, swing out the base curve into the pattern, and mark off the spacings a'',b''; b'',c''; c'',d''; d'',e''; e'',f'' equal to the corresponding spacings round the semicircle, and proceed to repeat them in the reverse order for the other half of the pattern. These spacing are not equal. From points a'',b'',c'',d'',e'' draw lines to the apex A. In accordance with right conic principles, project lines parallel to the base f,a' from the points a,b,c,d,e on the joint line to the outside slant of the cone. It so happens that these lines coincide with those of the cutting planes already there; thus, points $a,1,2,3,4,f$ are those required. Then, from these points, using the apex A as centre, swing arcs into the pattern to meet the corresponding radial lines from the base curve. Through the points of intersection draw the intersection curve as shown in the diagram. The remainder of the pattern is straight-forward right conic development and should present no difficulty.

The two smaller diagrams at (b) and (c) are given as extra examples for practice, since the lines of intersection may be obtained in precisely the same way.

OBLIQUE CONICAL HOPPER ON SPHERICAL DOME

The problem represented in Fig. 160 is that of an inverted oblique cone penetrating a spherical dome. The line of intersection occurs

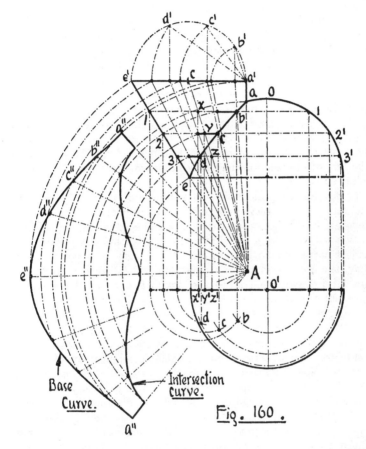

Base Curve.

Intersection Curve.

Fig. 160.

from a to e in the elevation, which must be determined before the pattern can be developed.

First complete the oblique cone by producing its sides a',a and e',e to meet at the apex A. Next draw its centreline C,A from the centre C of the base a',e' to the apex A. A number of cutting planes **may** now be drawn in any position. In the diagram three cutting planes are shown at $1,1'$; $2,2'$; $3,3'$, which cut the centreline of the oblique cone in points x,y,z. In this problem it is important to note that while the plan of the oblique cone at each cutting plane is a circle, the centres of these circles fall in different positions corresponding to the plan positions of x,y,z. These centres in the plan are shown at x',y',z'. The radii of these circles correspond to the distances $x,1$; $y,2$; $z,3$; in the elevation. Therefore, to obtain the plan circles of the oblique cone at the cutting planes, drop vertical lines from the points $1,2,3$ to the horizontal centreline in the plan, and describe arcs from the respective centres. Next, drop vertical lines from the points at the opposite ends of the cutting planes $1',2',3'$ to the horizontal centreline in the plan, and describe arcs, using O' as centre, to obtain the plan circles of the sphere at the cutting planes. The arcs from points $1,2,3$ cut the corresponding arcs from $1',2',3'$ in points b,c,d in the plan. Project these points vertically upwards to the respective cutting planes to obtain points b,c,d in the elevation. The line of intersection may now be drawn from point a, through points b,c,d to e.

To develop the pattern for the oblique cone, describe a semicircle a',e' on its base to serve as an inverted plan. From the apex A, draw elevation lines on the cone through points b,c,d on the line of intersection to the base a',e'. From the points obtained on the base, project perpendicular lines to meet the semicircle in points b',c',d'. Now, the apex A in the inverted plan will fall in the position a', and lines drawn from a' to the points b',c',d' on the semicircle represent the plan lines of those already drawn in the elevation through points b,c,d. In accordance with oblique cone development, from the plan apex a' swing arcs from points b',c',d', round to the base a',e', and then from the apex A, swing arcs from each of these points, together with arcs from the extreme points a' and e', into the pattern. Now, take from the semicircle the distances a',b' ; b',c' ; c',d' ; d',e' which are not equal, and, beginning at any convenient point a'' on the arc from a', mark off the distances a'',b'' ; b'',c'' ; c'',d'' ; d'',e'', stepping over from one line to the next. Proceed to repeat these spacings in the reverse order to point a'' at the other end of the pattern. A line drawn through these points from a'' to a'' will give the base curve in the pattern. Draw lines from each of the points on the base curve to the apex A.

The next step is to determine the intersection curve in the pattern. From the points on the base line a',e', which were obtained by describing arcs from points b',c',d', on the semicircle, draw lines to the apex A. These lines are the TRUE LENGTH lines of the elevation lines already there. Thus, the elevation line from point c' is that which passes through point c on the line of intersection, and its corresponding true length line is that drawn from the point on a',e', where the arc from c' falls. Now, the true length of the portion from A to c is found by taking the point c horizontally across to the true length line. The other true lengths of the portions A,b and A,d are obtained in the same way by projecting points b and d horizontally across to the corresponding true length line. In the diagram at Fig. 160, these horizontal connecting lines are shown in full black lines. Arcs from the points thus obtained on the true length lines should now be swung into the pattern, using the apex A as centre, together with arcs from the two extreme points a and e. Where these arcs meet the respective radial lines from the base curve, points will be obtained through which to draw the intersection curve in the pattern.

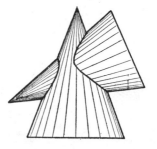

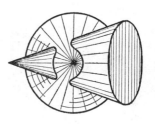

Fig. 161.

INTERSECTION OF TWO RIGHT CONES

The intersection of two right cones, a popular problem in advanced sheet metal geometry, falls to the method of cutting planes for the determination of the lines of intersection. Although, when the problem is presented in full, it appears to be somewhat complicated, the solution should offer little difficulty if the principles involved in the method of cutting planes and in right conic development be carefully borne in mind in following the working. The complete solution is best taken in two stages. First, the determination of the joint lines and, second, the development of the patterns. The illustration at Fig. 161 represents a vertical cone penetrated by an inverted cone somewhat smaller in size. This problem does not offer the simple solution obtained by the intersection of cones around a common central sphere, as dealt with in an earlier section.

As a preliminary to this problem, the method of obtaining the plan of a cone cut off at an angle, as illustrated in Fig. 162, will be dealt with first. The circle in the plan, representing the base of the cone, is divided into twelve equal parts, and the points 1' to 7' on the top half are projected vertically upwards to the base of the cone in the elevation. From the points 1 to 7 obtained on the base, lines are drawn to the apex A. These surface lines in the elevation are represented in the plan by joining the points on the circle to the centre A'. Now, where the cutting plane in the elevation crosses the

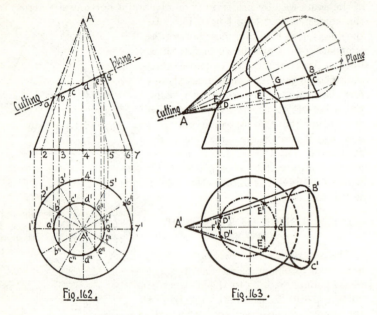

Fig. 162.   Fig. 163.

surface lines, the resulting points a,b,c,d,e,f,g should be dropped vertically downwards to cut the corresponding lines in the plan, as at a',b',c',d',e',f',g'. Similar points, $a'',b'',c'',d'',e'',f'',g''$, symmetrically opposite in the bottom half of the circle, will afford sufficient points to enable the full ellipse to be drawn in. The ellipse represents the plan of the cone at the cut-off.

In the problems of cutting planes previously considered, the cutting planes have been horizontal, or all parallel to each other. In this problem the cutting planes are neither horizontal nor parallel to each other. However, there is no difference in the method of solution, so that, except for the exercise of a little more care, no extra groundwork has to be considered.

Referring to Fig. 163, a semicircle is described on the base of the inverted cone in the elevation, divided into six equal parts, and the points projected perpendicularly back to the base. From the points on the base, lines are drawn to the apex A. These represent the usual lines drawn on the surface of a right cone, but in this problem it will be convenient to take these lines as representing the positions of cutting planes passing through both cones. For example, take the line AB, as shown heavily chain dotted in the elevation, and imagine that both cones are cut through at that plane. The plan of the inverted cone at the cut-off will be a triangle as shown at A',B',C' in the plan. The plan of the upright cone at the cut-off will be an ellipse as shown at F',G' in the plan. Where these two plan forms cut one another, as at D',E',E'',D'', points will be obtained which should be projected vertically upwards to the cutting plane in the elevation. Thus, points D and E are located, which lie on the intersection curves in the elevation. In order to obtain sufficient points on the intersection curves, this process should be repeated by taking cutting planes at the positions of the remaining lines on the inverted cone in the elevation. A series of ellipses will thereby be obtained in the plan, and these ellipses, taken at such angles as the cutting planes in this problem, are very nearly circles.

Fig. 164 shows the development of the patterns. As the intersecting bodies are right cones, where the radial surface lines cross the joint lines, the points are projected to the outside slant at right angles to the central axis of each cone respectively. Thus, taking the upright cone first, the radial surface lines are drawn from the base points 1,2,3,4,5,6,7 to the apex A. Where these lines, with the exception of $5A$, cross the intersection curves, the points are projected horizontally, or at right angles to the central axis, to the outside slant of the cone. From these points on the slant, and also from the base point 7, arcs are swung into the pattern. On the arc from the base, the spacings $1'',2'',3'',4'',5'',6'',7''$ are marked off equal to the distances $1',2',3',4',5',6',7'$ around the corresponding base circle in the plan. Radial lines are now drawn from the points on the base curve to the apex A, and where these radial lines meet the arcs from the slant side of the cone, points will be afforded through which to draw the joint curves in the pattern. Care must be taken, however, to see that the correct points are used in drawing the respective curves. Each point may be checked by tracing a connection from the point on the base line 1,7 to the corresponding point on the base curve in the pattern. For example, take point 3 on the base line, from there trace upwards along the surface line to the line of intersection, then horizontally across the cone to the slant side, around

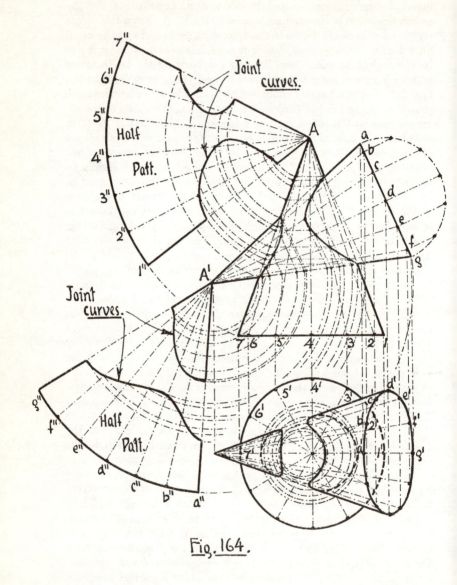

Fig. 164.

the arc to the joint curve in the pattern, and back along the radial line to point 3″ on the base curve.

Now taking the inverted cone, the radial surface lines are drawn from the points a,b,c,d,e,f,g to the apex A', and, where these lines cross the intersection curves, the points are projected AT RIGHT ANGLES to the central axis A',d to the outside slant of the cone. Note particularly that where the radial surface lines on this cone cross the intersection curves, the points do not occur at the same positions as do those on the upright cone, except by coincidence, or at such close proximity that the same point may be used. The procedure in developing the pattern is similar to that of the upright cone, and should be readily followed from the illustration given.

SIMILARITIES OF METHOD

It is sometimes surprising how a little alteration to the conditions of a problem will make it appear entirely different. The example

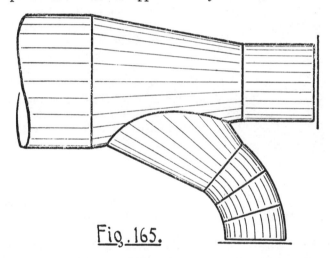

Fig. 165.

of the conical spout fitting on a conical body, as shown in Fig. 167, does not, at first sight, bear much resemblance to the conical inter-section given in Fig. 164; nor are the methods of obtaining the line of intersection usually compared. Yet the two problems are precisely the same, and the same method of finding the joint line may be applied in both cases. Since it is one of the features of this work to draw attention to similarities, the method of cutting planes, as applied to the solution of the conical intersection of Fig. 164, will be used in solving the spout problem shown in Fig. 167. The branch connection, sometimes used in pipe work as illustrated in Fig. 165, represents another example of conic intersection similar to the spout problem.

CONICAL SPOUT ON A CONICAL BODY

Referring to the elevation in Fig. 166, the spout cone AMN penetrates the vertical cone, or body, at the line of intersection,

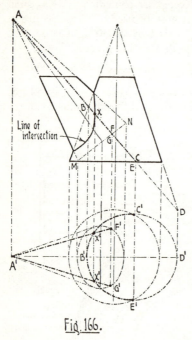

Fig. 166.

which must be determined before the patterns can be developed. In connection with this figure, the process of locating one point only, X, on the intersection is discussed.

Imagine that a cutting plane, along the line AD, cuts through both cones. The base of the vertical cone is produced from C to D in order to complete the cut through the cone. The plan of the vertical cone, from B to D, will be an ellipse, as shown at B',C',D',E'. The method of obtaining this ellipse in plan was given in connection with Fig. 162. The plan of the spout cone at the position AFG on this cutting plane will be a triangle as shown at $A'F'G'$. If the points in the plan at which this triangle cuts the ellipse, as in $X'X'$, be projected vertically upwards to meet the cutting plane AD, in the elevation, the resulting point X will lie on the line of intersection. This process may be repeated with any number of planes radiating from the apex A, cutting through both cones in order to obtain sufficient points through which to draw the line of intersection.

In the full problem shown in Fig. 167, a semicircle is described on the base MN of the spout cone, divided into six equal parts, and the points projected perpendicularly back to the base line. From the points on the base MN, lines are drawn to the apex A, and these lines are also produced backwards to meet the base of the vertical cone in points a,b,c,d,e. The positions of these lines are used as cutting planes passing through both cones, and the portions which cut the vertical cone are shown heavily chain dotted. In the plan, these cutting planes are shown as portions of ellipses, also heavily chain dotted, as seen at $a'a''$, $b'b''$, $c'c''$, $d'd''$, $e'e''$. The base MN of

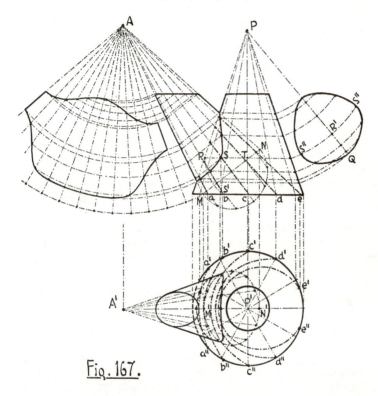

Fig. 167.

the spout cone is also projected into the plan as an ellipse, at $M'N'$, and the lines on the surface at the positions of the cutting planes are drawn from the ellipse to the apex A'. Where these lines on the spout cone in the plan cut the corresponding ellipses on the body cone, points will be obtained which should be projected vertically upwards to the respective cutting planes in the elevation. Thus, points will be obtained through which to draw the line of intersection.

In accordance with right conic development, project, at right angles to the central axis, all points on the line of intersection to the outside slant AM of the cone. Similarly, project all the points on the top edge of the spout to the outside slant AM. Next, from the apex A, swing out an arc from the base MN, and space off twelve divisions equal to those round the semicircle on MN. From the division points draw lines to the apex A. Next, from the apex A, swing arcs into the pattern from all the points on AM. Where these arcs cross the corresponding lines from the base arc, points will be afforded through which to draw the top and bottom curves in the pattern.

For the shape of the hole, project all the points on the line of intersection horizontally across the vertical cone to the opposite outside slant. From the apex P, swing arcs into the pattern from the points on the slant. Draw a centreline, PQ, in any convenient position. It remains now to determine the lengths of the arcs on either side of the centreline. These lengths should be equal to the horizontal distances round the vertical cone between the respective points on the joint line. To take one example, the length of the arc from R to S may be found by describing an arc, of radius RT, from the point T on the centreline of the cone, and dropping a vertical line from point S to cut the arc in S'. The length of the arc R,S' should now be marked off on both sides of PQ as at R',S''. This process should be repeated from each of the remaining points on the line of intersection, when the resulting points on either side of PQ should enable the contour of the hole to be drawn in.

AN ALTERNATIVE METHOD

The adaptability of the method of cutting planes is well illustrated by the alternative application shown in Fig. 168. Instead of taking cutting planes radiating from the apex A of the spout cone, which has the advantage of giving straight lines in the plan for that cone, horizontal cutting planes may be taken at any desired number of points between the extremities of the joint line. Horizontal cutting planes give a series of circles in the plan for the vertical cone, and ellipses for the spout cone. One of such planes is shown in the elevation as from B to F, Fig. 168. The plan of the vertical cone at the cutting plane presents the circle $C'F'$, while the plan of the spout cone gives the ellipse $B'E'$. The points at which these two figures cross, as at $X'X'$, when projected vertically upwards to the cutting plane BF in the elevation, locate one point, X, on the line of intersection. Two or three more points located in the same way by taking more cutting planes should afford sufficient to enable the line of intersection to be drawn in.

Although the full plan of the circle and ellipse is given for the sake of clearness, it may be worth while to look into the method of setting out, in the elevation only, sufficient of these details to solve the problem. Referring again to the elevation, Fig. 168, half of the ellipse is shown below the major axis BE. The minor axis of this ellipse may be found by making D the centre of BE. Then, through

D draw JG at right angles to the central axis AH of the spout cone. Next, with centre J and radius JG, describe the quadrant GH. Now, from D, draw DI parallel to JH, then DI is half of the minor axis of the ellipse. Therefore, from point D, draw DK at right angles to BE and equal in length to DI. The semi-ellipse BKE may now be drawn by any of the methods of drawing ellipses. The next step, by using L as centre and radius LC, is to draw the quadrant CM, which corresponds to $C'M'$ in the plan. The quadrant CM cuts the semi-ellipse in point X'', which may now be projected vertically upwards to the cutting plane BF, thus locating point X on the line of intersection.

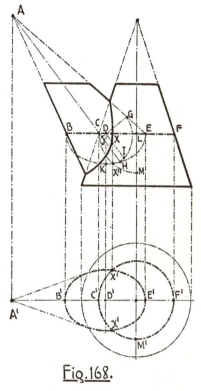

Fig. 168.

Whether the full plan method is adopted, or whether the solution is worked on the elevation only, the total amount of work involved will be much the same. Therefore, it is advisable to use the method which presents itself most clearly.

TALLBOY TRANSFORMER INTERSECTED BY CYLINDRICAL PIPE

There is yet another type of intersection which can be successfully solved by the method of cutting planes. Fig. 169 shows a tallboy transformer intersected by a cylindrical pipe. The joint line may be accurately determined by the method of cutting planes, as illustrated in Fig. 170.

Assuming that the main particulars of the plan and elevation have been set down to the required dimensions, the first step towards obtaining the line of intersection lies in allocating the positions of the cutting planes. In Fig. 170, a semicircle is described on the end

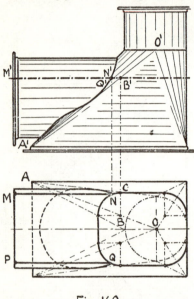

of the cylinder, both in the plan and elevation. This, in each case, is divided into six equal parts, and in the elevation the points are numbered 1,2,3,4,5,6,7. In the plan, since the view is looked at from above, the points around the circle are numbered as shown at 1',2',3',4',5',6',7'. Thus, points 1' and 7' fall on the centreline, which corresponds to the top and bottom lines 1 to 7 in the elevation. From the points on the semicircle in the elevation, lines 2,2; 3,3; 4,4; 5,5; 6,6 are drawn horizontally right through the cylinder and transformer to represent the positions of cutting planes. The advantage of taking cutting planes at these

Fig. 169 .

positions will be evident from the fact that the plan of the cylinder cut through at each of these planes will be two parallel straight lines from the corresponding points on the semicircle in the plan. The plan of the transformer at these positions, however, is not quite so simple. Nevertheless, the observance of a few straightforward principles should help to keep the solution clear.

Referring now to the plan, Fig. 170, the circle representing the top of the transformer is divided into four quadrants. Each of those four quadrants is joined to the corresponding corner at the base of the transformer, forming four corner pieces with flat triangular faces between them. Each of these corner pieces is a quarter of an oblique cone. Therefore, in cutting these through horizontally at the position of one cutting plane, four equal quadrants are formed, of smaller radii than those forming the top circle, and the quadrants are joined by straight lines between them. This is shown clearly in the plan at Fig. 169. Now, the centres of these quadrants can easily be located, since each centre must lie on a line joining the centre O, of the top circle to the respective corner of the base. One of these centrelines is lettered OA in the plan at Fig. 169. This line in the elevation

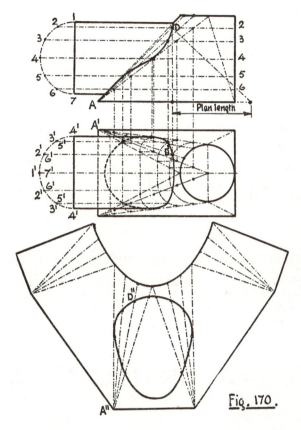

Fig. 170.

occurs at $A'O'$, and the point B', at which the cutting plane crosses, may be dropped into the plan to obtain B. Then BC becomes the radius of the quadrant. Similar quadrants may be located in each of the other four corners, and at the same cutting plane the four quadrants will have equal radii. Next, in the plan at Fig. 169, the width of the cylinder at the cutting plane is indicated by the two parallel lines MN and PQ. These two lines cut the quadrants in N

and Q respectively, and by projecting those two points vertically upwards to the cutting plane in the elevation, the point N', which coincides with Q', is obtained on the line of intersection.

Referring back to Fig. 170, this process is repeated at each of the cutting planes 2,2; 3,3; 4,4; 5,5; 6,6, by means of which sufficient points are obtained to enable the line of intersection to be drawn in. More cutting planes will, of course, afford more points and ensure greater accuracy if desired. The pattern for the transformer is shown with the contour of the hole for the cylinder. The development of the transformer pattern is straightforward triangulation, and has been fully dealt with in an earlier stage. Triangulation is also used to obtain the hole in the pattern, and the method of obtaining one point is here described. The point marked D' in the plan, Fig. 170, is that in which the line $2'D'$ intersects the quadrant on the cutting plane 2,2. The plan length $A'D'$ is taken and marked off along the base in the elevation at right angles to the vertical height up to point D. The diagonal then gives the true length of AD, which should be marked off in the pattern along the corresponding line, as from A'' to D''. This process, repeated with the other points of intersection, will give sufficient points through which to draw the contour of the hole.

CYLINDER PENETRATING TRANSFORMER AT AN ANGLE

The problem at Fig. 170 shows the cylinder symmetrically placed on the horizontal centreline of the transformer. Had the cylinder penetrated at an angle such as shown in Fig. 171, the problem would have every appearance of being more difficult to solve. But this is not so. Exactly the same method is used, as careful inspection of the details will show, and except for the exercise of a little more care, no extra work or fresh principles are involved. In this example the pattern for the cylinder is shown developed, which can very easily be "unrolled" in the usual way when the line of intersection is obtained in the plan.

INTERSECTION OF TALLBOY TRANSFORMER AND RIGHT CONE

Although the general principles of solution are the same, there are one or two points of difference in the problem shown in Fig. 172, which will make a little investigation worth while. In the first place, the top of the transformer is inclined at an angle to its base, and in taking horizontal cutting planes the corners at each plane in the plan

will be quarter-ellipses instead of quarter-circles. The plan of the
cone at each plane will be a circle.

In the problem at Fig. 172, four cutting planes are taken at
A,B,C,D. To obtain the plan of the cone at these positions, drop the
points A,B,C,D vertically downwards to the horizontal centreline
in the plan, and with centre O, describe a circle from each point. To

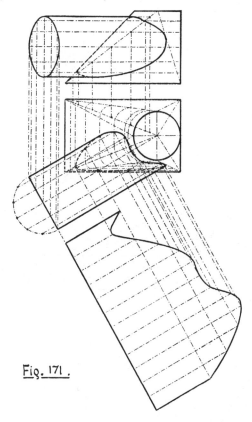

Fig. 171.

obtain the plan of the transformer at the cutting planes, describe
a semicircle on the top edge of the transformer and divide it into six
equal parts, as from 1 to 7. Project these points perpendicularly
back to the top edge 1,7, and from there drop them vertically downwards into the plan to obtain the ellipse 1' to 7'. Join the points
4',5',6',7' to the corner M' in the base on both sides of the centreline. Next, in the elevation, join the points 4,5,6,7 on the top edge to

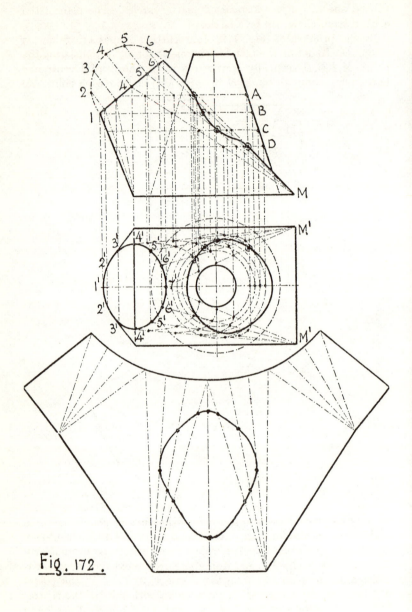

Fig. 172.

the equivalent corner M. Now from the points at which the cutting planes A,B,C,D cut these lines, drop verticals to cut the corresponding lines in the plan. Points will thereby be obtained through which to plot the quarter-ellipses, which represent the shape of the corners at the positions of the cutting plane. Next, locate the points where, at each cutting plane, the plan of the cone cuts the plan of the transformer. These points in the illustration at Fig. 172 are marked in the plan with small circles round them in order to distinguish them more clearly. Now, project these points vertically upwards to the corresponding cutting planes in the elevation, where they will give points on the line of intersection. With care, this line may now be drawn in.

The contour of the hole in the pattern of the transformer is shown in Fig. 172. This may be determined by locating points on the corner lines of the transformer in precisely the same way as described in connection with Fig. 170. The pattern for the right conical connection may be developed by the radial line method, as described in an earlier course, and is quite straightforward when the line of intersection is determined.

OBLIQUE CONICAL HOOD INTERSECTING VERTICAL FLUE

The application of the method of cutting planes to problems of the oblique cone does not involve fresh or more difficult principles than those used in solving intersections of the right cone. The base of an oblique cone is a circle, also any cross-section parallel to the base is a circle, as at DE or FG, Fig. 175. Therefore, when the base of the oblique cone is horizontal a series of horizontal cutting planes will present a corresponding series of circles in the plan, as shown in Fig. 175.

The intersection of the oblique conical hood and the vertical flue, shown in Fig. 173, is usually dealt with by means of a series of radial lines from the apex of the cone, but the line of intersection may be as readily found by the application of cutting planes. This alternative method is given in Fig. 173.

In this case three horizontal cutting planes are taken, as at DE, FG, HI, in any convenient positions. Any number of cutting planes may, of course, be taken to suit circumstances, but a large number of cutting planes does not necessarily lead to a high degree of accuracy in plotting the shape of the intersection line. Sometimes a few well-placed points will serve to determine the position of the curve, when a little careful drawing is all that is necessary to produce a good line of intersection.

To locate the line of intersection, drop points D,F,H vertically downwards to the base *BC*. Also, drop the points *d,f,h* vertically downwards to *d',f',h'* on the base line *BCA'*. The latter points are the centres of the circles *D,F,H*. Therefore, from points *d',f',h'*, with the corresponding radii, draw arcs to cut the plan of the cylinder in points *D',F',H'*. From points *D',F',H'* project lines vertically upwards to points *D",F",H"* on the corresponding cutting planes in the elevation. A curve drawn from *M*, through these points to *N*, should give the required line of intersection.

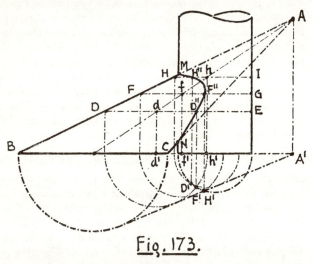

Fig. 173.

OBLIQUE CONE INTERSECTING A RIGHT CONE

A further example of oblique conic intersection is given in Fig. 174, where an oblique cone penetrates a right cone with the axes off-centre. Assuming that the plan and elevation of the two cones are set down to required particulars, with the off-centre position of the oblique cone as shown in the plan, then the first step towards obtaining the line of intersection is to locate a number of cutting planes in convenient positions, as shown in the elevation at *DE*, *FG*, *HI*, *JK*, *LM*. Drop the points *D,F,H,J,L* vertically downwards to the horizontal centreline of the right cone in the plan, and from each of the points obtained on the centreline describe circles, using *A'* as centre, to represent the plan of the right cone at each of the cutting planes. In the diagram at Fig. 174 three-quarters only of the circles are drawn. Next drop the points *d,f,h,j,l* on the centreline of the oblique cone vertically downwards to the corresponding centreline in the plan. Now drop the remaining points E,G,I,K,M, vertically downwards to the centreline of the oblique cone in the plan, and with radii dE, fG, hI, jK, lM describe circles to represent the plan of the oblique cone at each of the cutting planes. Where the corresponding pairs of circles cut each other, that is, where the right conic circles cut the oblique conic circles at their respective

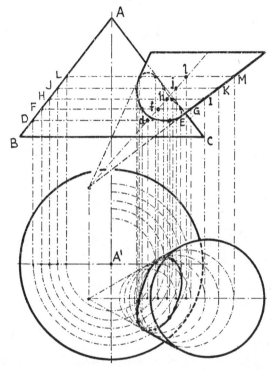

Fig. 174.

planes, points will be afforded in the plan through which to plot the intersection curve. To obtain the line of intersection in the elevation, the points in the plan should be projected vertically upwards to the corresponding planes in the elevation. The required curve may then be drawn in as shown in the diagram.

OBLIQUE CONIC SECTIONS

It may sometimes be required to obtain the shape of a cross-section of an oblique cone, as, for example, when a baffle or valve is to fit across an oblique conic transition piece. Before explaining the processes of obtaining the shape of any cross-section, it may be well to revise one or two properties of the oblique cone which have an important bearing on the methods in view.

Attention has already been drawn to the fact that any cross-section parallel to the base is a circle. Referring to Fig. 176, the line of centres OA passes through the centre of any such circle. Thus, the point O', on OA, is the centre of the circular cross-section EF. Nevertheless, the line of centres OA does not coincide with the central

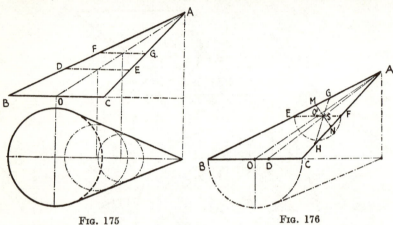

FIG. 175   FIG. 176

axis of the cone. The position of the central axis may be found by bisecting the angle BAC, which gives the line AD. The central axis is that line around which the body of the cone is symmetrically formed. Thus, when AM equals AN, the line joining MN will be bisected in S by the central axis AD. The shape of the cross-section at MN, however, is an ellipse, and not a circle. Moreover, any cross-section which is not parallel to the plane of the base is an ellipse, except when parallel to the subcontrary section. The subcontrary section in relation to EF is given at GH. The line GH is equal to EF, and passes through S in such a position that SG equals SF, and SE equals SH. Then the shape of the cross-section at GH again becomes a circle.

ELLIPTICAL CROSS-SECTION

There are two useful methods of finding the shape of an elliptical cross-section of an oblique cone, the choice of which will depend a good deal on the purpose in view. For example, the method shown in Fig. 177 is useful in surface development, while the method given in Fig. 178 will be found sufficient when the shape of the cross-section only is required, as for a baffle plate.

Referring to Fig. 177, to find the shape of the cross-section at CD, first describe a semicircle on the base BC, divide into six equal parts, and join the points to the plan apex A'. Next, project the points on the semicircle perpendicularly up to the base line BC, and from the points on the base line draw elevation lines to the apex A. The points 1,2,3,4,5 where the elevation lines cut the cross-section CD may now be dropped vertically downwards to the corresponding plan lines to obtain the points $1',2',3',4',5'$. Draw in the plan of the cross-section. The next step is to project lines from points 1,2,3,4,5

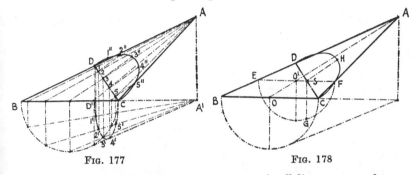

Fig. 177   Fig. 178

at right angles to CD, and on these lines mark off distances equal to the corresponding perpendicular distances from $D'C$ in the plan. Thus, make $1,1''$ equal to the perpendicular distance from $D'C$ to $1'$, and make $2,2''$ equal to the perpendicular distance from $D'C$ to $2'$, and so on. When this is completed, draw in the semi-ellipse on CD. This represents half of the true shape of the cross-section at CD.

For the alternative method refer now to Fig. 178. The line CD represents the minor axis of the elliptical cross-section, and SH half of the major axis. First locate S, the centre-point of CD. Then, through S, draw EF parallel to the base BC. Thus, the cross-section at EF is a circle. Next locate O', the centre-point of EF, and describe the semicircle thereon. Now, to find the width of the cone through point S at right angles to the plane of the diagram, drop a line from S perpendicular to EF to cut the semicircle in point G. Thus SG is half of the required width of the cone, and represents half of the major axis of the ellipse. Therefore, project SH at right angles to CD and equal in length to SG. Then the semi-ellipse, CHD, may be drawn by any of the methods adopted for drawing ellipses and ovals.

CHAPTER 12

BRANCH AND JUNCTION PIECES

GEOMETRICAL obscurities, or little differences not easily seen, are
sometimes responsible for perplexing errors. The pattern for the
two-way breeches piece shown in Fig. 179 (a), with branches of equal

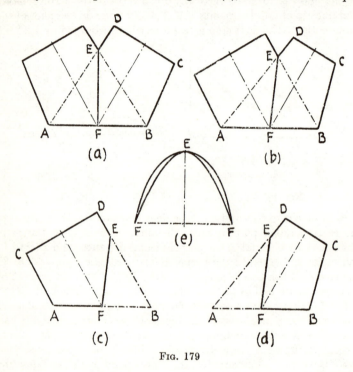

FIG. 179

diameters, may be easily and correctly developed by taking one limb
only, *ABCD*, between the top and bottom circles, and developing
the pattern for the portion *BCDEF*, with the joint line at *EF*. Fig.
179 (b) shows a similar breeches piece, but with branches of different
diameters. At first sight there appears to be no reason why the
patterns for the two limbs should not be developed separately, as
shown below at (c) and (d), with the joint line at *EF*, but if this were

236

done the joint at EF would not fit. Although the points E and F coincide on each branch, the curves themselves do not match. The figure at (e) shows the difference between the shapes of the curves as they would occur at EF in the figures at (c) and (d). The larger or outer curve corresponds to that in (c), and the smaller inner curve would be obtained in the figure at (d).

These curves are geometrically true, but it will be readily seen that in the shaping up of the respective branches in practice it would be an easy matter to make the shape of the inner curve correspond to the form of the outer. Nevertheless, if the patterns were developed carefully on these lines, the length of the outer curve would be longer than the inner, thus making the actual fitting somewhat troublesome.

BREECHES PIECE OF UNEQUAL DIAMETERS

This brings out an important point in developing patterns for breeches pieces with branches of unequal diameters. Referring now to Fig. 180, the larger limb is developed as previously suggested by joining the points on the top circle to the corresponding points on the bottom circle. Thus, the lines in the elevation from points $7',9',11',13'$ are drawn through to meet the equivalent points in the base. These lines pass through the points $8',10',12',14'$ on the joint line. Now, if the lines from points $9'$ and $11'$ on the smaller limb were drawn through to meet their equivalent points on the base, they would not pass through points $10'$ and $12'$ on the joint line. This may possibly be seen to greater advantage in the plan. Taking, for example, the line from point 9 on the top edge of the larger limb, this passes through point 10 on the joint curve to point 6 on the base. Now, taking the line from the corresponding point 9 on the top edge of the smaller limb, if this were drawn through to the equivalent point 6 on the base, it would miss point 10 by a considerable amount, as will be seen in the illustration. Therefore, the lines from points 9 and 11 on the smaller limb are diverted to meet points 10 and 12 in the joint curve. By doing this, and developing the pattern accordingly, the inaccuracy of the joint curves will be overcome.

To develop the patterns, first set out the elevation of the two limbs, and on the top edge, $1'$ to $13'$, in each case, describe a semicircle, divide into six equal parts and project the points perpendicularly back to the respective edges. Drop these points vertically downwards into the plan and plot the semi-ellipses 1 to 13. Divide the base semicircle into six equal parts, as shown at 2,4; 4,6; 6,8; 8,6; 6,4; 4,2, and project these points vertically upward to the base line 2,2. Now, on the larger limb only, join the corresponding points on the top and bottom edges, both in the plan and elevation. Thus, in the plan 11 will be joined to 4, 9 to 6, 7 to 8, and so on, and in the elevation, 11′ to 4′, 9′ to 6′, 7′ to 8′. The lines 9′,6′ and 11′,4′ in the elevation cross the joint line in points 10′ and 12′. Drop these points,

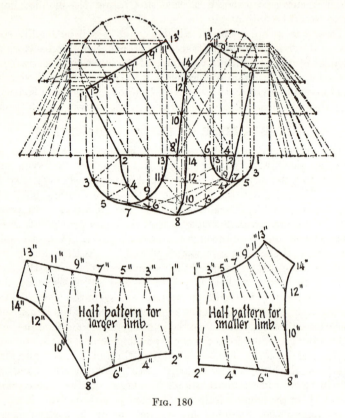

FIG. 180

with point 14′, vertically downwards to obtain the points 10, 12, and 14, in the plan. The joint curve 8 to 14 may now be drawn in, and on the smaller limb the points 9 and 11 may be joined to points 10 and 12 of the joint curve.

In the illustration at Fig. 180, a half-pattern is given for the larger limb, and a half for the smaller. For the larger pattern, the first line 1″, 2″ may be taken direct from the elevation, as from 1′ to 2. Next take the plan length 2,3, and triangulate this against the appropriate vertical height on the corresponding vertical height line.

Take the true length line and from point 2″ in the pattern swing an arc through point 3″. To complete the first triangle the true distance 1″,3″ should be taken from the semicircle on the top edge 1′,13′. Therefore, take one of the equal spacings, and from point 1″ in the pattern, describe an arc cutting the previous arc in point 3″. For the next triangle take the plan length 3,4, and triangulate it against the appropriate vertical height. Take the true length diagonal and from point 3″ in the pattern swing an arc through point 4″. The next true length may be obtained direct from the plan. Thus, take the true plan length 2,4, and from point 2″ in the pattern describe an arc cutting the previous arc in point 4″. This process may be repeated as far as the line 8″,9″. Then, for the next triangle, take the plan length 9,10, and triangulate it against its vertical height, this time using the base line level with point 10′. Take the true length diagonal and from 9″ in the pattern swing an arc through point 10″.

The next step is an important one. Up to this stage the true distances 2″,4″; 4″,6″; 6″,8″ have been taken direct from the base circle in the plan, but the next distance in the plan, 8 to 10, is not a true length, since it is the first spacing up the joint curve. Therefore, it should be triangulated against its vertical height, by taking the plan length 8,10, marking it off along the bottom base line and taking the true length diagonal up to the point level with 10′ on the vertical height line. With this true length in the compasses, from point 8″ in the pattern, describe an arc cutting the previous arc in point 10″. The remainder of the pattern may now be completed, taking care to triangulate the plan lengths 10,12 and 12,14 against their respective vertical heights in order to obtain the true distances.

The pattern for the smaller limb may be developed in the same way, and with the caution that the true distances 8″,10″; 10″,12″; 12″,14″ must be the same for both patterns, the smaller example is left to be completed without further explanation.

FOUR-WAY BREECHES OF UNEQUAL BRANCHES

The illustration in Fig. 181 shows an application of a four-way branch piece of similar construction to the two-way example dealt with in Fig. 180. It will be noticed in this case, however, that the top of the branch piece is horizontal, conforming to the centreline of the two larger bends which turn into reverse directions at right angles to the original flow. The two lower branches, smaller in diameter, are directed to positions vertically below. The problem of the pattern developments is given in Fig. 182, and, while the principles involved are the same as in the two-way breeches piece, the off-centre positions introduce some measure of intricacy which may

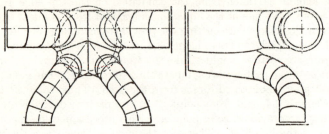

FIG. 181

require a little more careful attention. Therefore, only the two limbs on the left are developed for patterns. The elevation above the plan

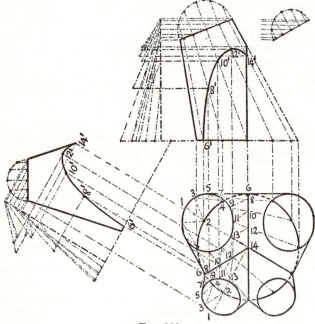

FIG. 182

is of the larger limb only. The elevation of the smaller limb is projected at right angles to its centreline in order to make the solution

clearer, and to get a direct angle of the top edge. Since the larger
limb is not symmetrical about any axis, the full pattern must be
developed.

The method of dividing the surface into triangles is similar to that
of the previous problem. In the four-way example, the numbering
on the larger limb begins in the middle of the back, and proceeds
round the surface to the seam at 13,14. The other half is divided
similarly, but the points are not numbered in order to avoid undue
congestion. The smaller limb is symmetrical about its centreline,
and one half only, that
which matches up to the
larger limb, is divided
into triangles. The
points 6,8,10,12,14 on the
joint line between the
two limbs, are common to
both. In setting out the
projected elevation of the
smaller limb, it is therefore necessary to mark
off points 8′,10′,12′,14′ to
the same vertical heights
above the base line as
the corresponding points
in the larger elevation.

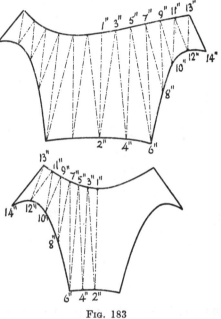

As these points are the
same vertical height in
each case, it may be
observed that, for the
purposes of pattern de-
velopment, the projected
view of the smaller limb

FIG. 183

may be dispensed with provided that the angle of the top edge is
set out in the larger elevation in such a position that the points
on it may be projected horizontally to the vertical height line in
that view. Then, with a little care, all the true lengths necessary
for both patterns may be obtained from the one vertical height
line. The directions for drafting the pattern are almost a repetition
of those for the two-way breeches of Fig. 180, and if the explanation
has been carefully followed, no difficulty should be experienced
in obtaining the patterns for the four-way breeches. The patterns
are shown in Fig. 183.

A FOUR-WAY JUNCTION PIECE

There are many ways in which a multiple branch piece might be made, the design and method of construction depending on requirements and circumstances. It often happens that branches are required to be taken from a duct in a limited space, in which two or three branches are sometimes needed close together. With a little careful design, a single branch piece may be made to incorporate all that is required in the matter of transforming the shape and adjusting the areas.

The branch piece shown in Fig. 184 is a junction of three branches and a main, comprising four ways in all. The two side branches turn

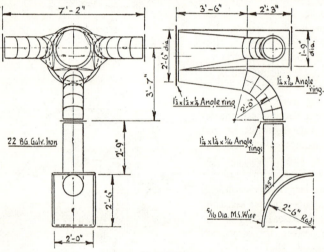

FIG. 184

into the horizontal at right angles to the main, and the third branch turns vertically downwards to a hood below. The main is reduced in diameter at the branch end to adjust the area simultaneously with the division of the air flow. This junction piece belongs to the type with small flat triangles between the branches, where the metal is bent in the folders and the form shaped up on the bar. Fig. 186 shows a half-pattern which is symmetrical about the centreline at 1',2'. This four-way junction piece can, if the dimensions are not too large, be made in one piece of metal with the same either at 1,2 or 34,35. It is a simple method of construction and saves a lot of work when the form and mode of structure are clearly understood.

The diagram at Fig. 185 is of the junction piece only, showing the holes for the branches inclined at an angle of 15 degrees, thus leaving the bends to be made up of five segments, each of 15 degrees, to complete the 90 degree branches. The pattern-development is obtained by straightforward triangulation, and as the branch piece is symmetrical about the centreline 2 to 34, only one half of the pattern is shown

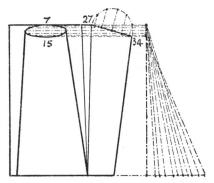

developed. The method of dividing the surface into triangles is extremely important, since the lines forming the triangles must lie on the surface of metal and not cut across space. Sometimes in triangulation this cannot be entirely avoided, since on a conical surface the alternate lines forming the triangles must lie slightly off the surface of metal, but the arrangement of the triangles must be such that this fault is reduced to a minimum. For example, in Fig. 185 it will be seen that the straight line between points 2 on the bottom and 3 on the top must lie slightly across or inside the surface. On the other hand, the lines radiating from point 16, as, for instance, lines 16,18; 16,19; and so on from 16 to points 20,21,22 . . . 30,31,32, all lie perfectly on

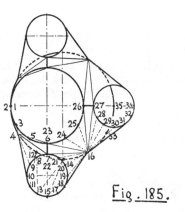

Fig. 185.

the surface of the metal. When this condition is fulfilled there should be no difficulty in obtaining a correct pattern.

The large semicircle of the base is divided into six equal parts, which in the diagram bear the numbers 2,4,12,14,16,33,35. The smaller semicircle of the top is also divided into six equal parts, the points of which are numbered 1,3,5,6,24,25,26. The circular edge of the lower branch is divided into twelve equal parts, as at points 7,8,9, . . . 20,21,22, and the semicircle of the middle branch is divided into six equal parts, as at points 27,28,29,30,31,32,34. The various points are joined up as shown in the diagram in accordance with the prescribed method of numbering. The plan is so placed that the middle branch 27,34 is on the horizontal centreline, thus giving in the elevation the true angle of its slope. The tops of the other two branches are inclined at the same angle, so that the corresponding points on those circles will have the same vertical heights. Thus, point 15 on the lower circle has the same vertical height from the base as point 34. Similarly, points 10 and 19 have the same vertical height as point 30. Bearing this in mind, it will be seen that the points on the top edge 27,34 in the elevation, projected horizontally to the vertical height line, will also serve as vertical heights for the points on the circular edge 7,15.

For the pattern, the first line 1',2' may be taken direct from the elevation since it is equal to the full vertical height, and marked off in any convenient position. Next take the plan length 2,3, and mark it off along the base line at right angles to the vertical height. Take the true length diagonal up to the top and from point 2' in the pattern swing an arc through point 3'. Now take the true distance 1,3 direct from the plan, and from point 1' in the pattern describe an arc cutting the previous arc in point 3'. For the next triangle take the plan length 3,4, and mark it off along the base line at right angles to the vertical height. Take the true length diagonal up to the top and from point 3' in the pattern swing an arc through point 4'. Next take the true distance 2,4 direct from the plan and from point 2' in the pattern describe an arc cutting the previous arc in point 4'. Repeat this process with plan lengths, 4,5; 4,6; 4,7, triangulating each against the full vertical height, and for the top true lengths take 3,5; 5,6; 6,7 direct from the plan. For the next triangle take the plan length 4,8, and mark it off at right angles to the vertical height line. Take the true length diagonal, this time up to the first point below the top, which corresponds to the vertical height of point 8, and from point 4' in the pattern swing an arc through point 8'. Next, the true distance between 7 and 8 should be taken from the semicircle in the elevation, because in the plan the top edges of the branches are really ellipses, and do not give the true distances between the points thereon. It should be noted, however, that in this case, as the top edge is only inclined at 15 degrees, the plan ellipses are very nearly circles. When the top edges are inclined at, say, 30 degrees or more, the plan ellipses are correspondingly much more pronounced.

For the next triangle the plan length 4,9 should be triangulated against the vertical height, and the true length diagonal taken, this time up to the second point down from the top. The triangle in the pattern should then be obtained as before. This process of triangulation should be repeated with plan lengths 4,10 and 4,11, which completes all the radiating lines from point 4. The next plan line passes from point 11 in the top to point 12 on the bottom semicircle, and the plan lines following it, 12,13; 13,14; 14,15; 15,16, alternate between the top and bottom. On reaching point 15 it should be noted that the lowest point on the vertical height line has been

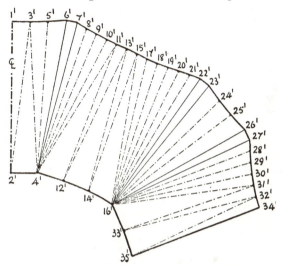

Fig. 186.

arrived at, and in triangulating the next series of plan lengths radiating from point 16 in the bottom to 17,18,19,20,21,22 in the top, the vertical heights step up again to the top. In triangulating the next series of lines radiating from point 16 to points 23,24,25,26,27, the vertical height remains at the top. In the next series, from 16 to points 28,29,30,31,32, and then from 32 to 33 and 33 to 34, the vertical heights again step down to the lowest point. With care in following these observations the full pattern should not present any real difficulty in completing.

A THREE-WAY JUNCTION TRANSFORMER

The junction piece presented in Fig. 187 is of similar construction to that of the previous problem, except that it transforms from a rectangle to three circles, and the two smaller branch circles incline at 30 degrees instead of 15 degrees. The latter condition makes the

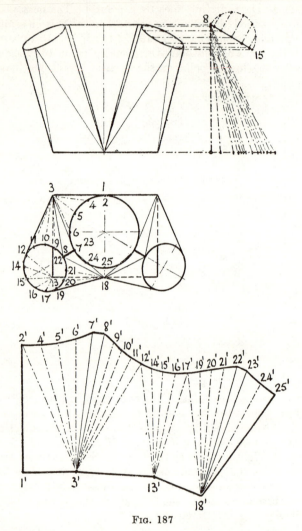

FIG. 187

plan of the branches more elliptical than those of the problem shown in Fig. 185. An important diversion in dealing with this problem occurs in connection with the vertical heights of the points on the branches. It will be noted that in this case neither of the branches in the plan occurs on a horizontal centreline, thereby failing to give a true engle of slope of the top edge in the elevation. Since the top edge of both branches is known to incline at 30 degrees, all that need be done is to draw a line at 30 degrees from the top of the vertical height line equal in length to the diameter of the branch, as shown at 8,15. Describe the semicircle on it, divide into six equal parts and project the points perpendicularly back to the line. From the points obtained on this line project horizontals back to the vertical height line. These points then give the required vertical heights of the points 8,9,10, . . . 20,21,22, around the ellipse in the plan.

With due care in dividing the surface of the junction piece as shown in the diagram, and following the method of triangulation as described in the previous problem, the pattern should now be readily drafted.

Chapter 13

THE TAPERED LOBSTER-BACK

COMPARE the two lobster-back bends shown in Fig. 188. One is an ordinary six segment 90 degree bend, with diameters equal at each end. The other is a tapered six segment 90 degree bend, with diameters different at each end. There are several methods of making tapered lobster-back bends, the design and construction depending a good deal on the purpose in view.

TAPERED LOBSTER-BACK FROM OBLIQUE CONE

For a tapered lobster similar to that shown in Fig. 188, the patterns may be developed as segments of an oblique cone. Where no allowances on the patterns are required, as in the case of welded seams, this method has a number of advantages, among which are a minimum of scrap and a minimum of cutting.

The illustration at Fig. 188 (a) shows that the bottom segment in an ordinary non-tapered bend is a part of an oblique cylinder, and the corresponding diagram below, at (c), shows how the cylinder, cut into six segments, may be made to form the bend when the alternate segments 2,4,6 are turned round through 180 degrees. Since each segment includes an angle of 15 degrees, the centreline CL of the bottom segment makes an angle of $7\frac{1}{2}$ degrees with the base line. As the central axis of the cylinder is at right angles to the centreline CL of the segment, the central axis of the cylinder must, therefore, lean at an angle of $7\frac{1}{2}$ degrees from the vertical. This is an important point, since, in the case of the tapered lobster-back, the same line of reasoning applies to the oblique cone. Each segment of the tapered lobster-back at (b) contains an angle of 15 degrees. The centreline CL of the bottom segment is at $7\frac{1}{2}$ degrees with the base line. Therefore, the centreline of the oblique cone must incline at $7\frac{1}{2}$ degrees from the vertical.

There is one important difference between the ordinary bend and the tapered bend. In the former case, the centreline of the bend lies on a quarter-circle with the centre at O, see Fig. 188 (a). In the latter case, the centreline of the tapered bend takes the form of an ellipse. The reason for this will be seen by an inspection of Fig. 189. The oblique cone, 1,A,2, contains the bottom segment of the bend at 1,2,4,3, with the centreline passing through a,b. Although point a is the centre of the base 1,2, it will be found that the corresponding point *b* does not lie at the centre of 3,4. Now, if the top portion of the cone, 3,*A*,4, be turned over, through 180 degrees, so that the point 3 falls on point 4, and point 4 falls on point 3, the apex *A* will move to *B*, and the centreline *bA* will move to *b'B*. The point *b*,

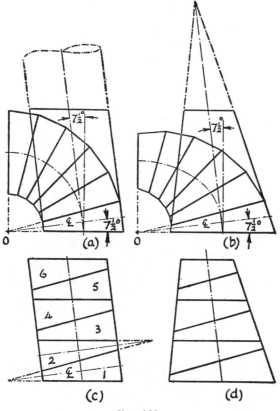

Fig. 188

therefore, has moved to point *b'*.   The next segment now lies at 3,4,6,5, with the centreline passing through *b'c*. If the reversing process be again applied to the top portion, 5,*B*,6, so that the point 5 falls on point 6, and point 6 falls on point 5, the apex *B* will move to *C*, and the centreline *cB* will move to *cC*. This time, however, it will be found that the centrepoint *c* remains at the same spot. The reason for this is because the line 5,6, in the cone before reversal, is parallel to the base line 1,2, which results in the centreline, aA, passing through the centre-point c of 5,6. Following up this process of reversal of the segments, it will be seen, in Fig. 189, that the centre-point of each alternate joint line falls farther from the true quadrant. Therefore, if the positions of the two diameters at each end of the bend are to be accurately placed to satisfy a particular design, this discrepancy, which varies with the amount of taper, must be carefully checked beforehand. When the accuracy of these positions is not important, this method may be found very useful.

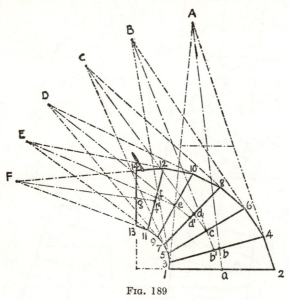

FIG. 189

The development of the patterns is shown in Fig. 190. First describe a quadrant approximating to the required centre radius of the bend, and divide it into six equal parts. Erect a centreline O,A for the cone at an angle of $7\frac{1}{2}$ degrees from the vertical, and mark off from its base upwards the six divisions from the quadrant. Through the top and bottom points draw horizontal lines and mark off the large diameter at the bottom and the small diameter at the top. Complete the cone by drawing the outside lines through the extremities of the diameters. These outside lines should meet at the apex A on the centreline. Through the remaining division points draw the joint lines alternately at 15 degrees, as shown in the diagram, to represent the segments.

To develop the patterns, drop the apex A vertically downwards to A' on the base line. Describe a semicircle on the base line and divide it into six equal parts, as shown at 1,2,3,4,5,6,7. From the apex A' in the plan, swing the points 1 to 7 on the semicircle round to the base line. From the apex A in the elevation swing the points

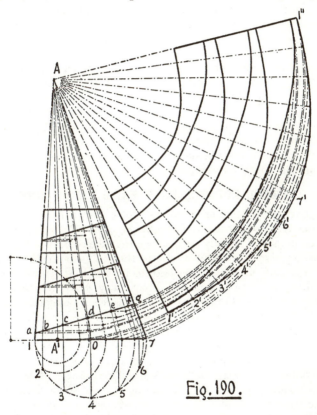

Fig. 190.

obtained on the base line round into the pattern. Take one of the equal divisions from the semicircle, as from 1 to 2, and, beginning on the inside arc, space off the base curve $1'$ to $7'$ to $1''$, in the pattern. Join the points on the base curve to the apex A. Next project the points on the semicircle vertically upwards to the base 1,7, and draw lines from the points obtained on the base to the apex A. These lines are ELEVATION lines. Next draw the TRUE LENGTH lines by joining to the apex A the points on the base 1,7, which were obtained by swinging the points 1,2,3,4,5,8 round from the plan apex A'. In the illustration at Fig. 190, these true length lines are shown chain-dotted while the elevation lines are shown in full. Now, taking the first joint line a,g up from the base, the elevation lines cross the joint line in points b,c,d,e,f. These points, in accordance with oblique conic development, are projected horizontally to meet the corresponding true length lines. For example, the elevation line from point 3 crosses the joint line in point c. The corresponding true length line is that obtained by swinging point 3 round to the base and then drawing a line from the point on the base to the apex. Therefore, point c is projected horizontally across to meet this line, and from there it is swung into the pattern to meet the radial line $A,3'$. This process, repeated with the other points, a,b,d,e,f, should give points in the pattern through which the joint curve may be drawn. Each of the other curves in the pattern may be plotted in the same way from the points where the elevation lines and true length lines cross the joint lines in the elevation.

TAPERED LOBSTER-BACK FROM RIGHT CONE

A slight modification of the above problem may transform it into one of a right cone instead of an oblique cone. Referring again to Fig. 188 (*b*), it has already been shown that the centreline CL of the bottom segment is at right angles to the centreline of the oblique cone. Imagine, now, that the bottom part of that segment below the centreline CL to be removed or cut away, and the lobster-back allowed to fall back so that that centreline becomes the base. Thus, the lobster-back falls through $7\frac{1}{2}$ degrees and the centreline of the cone becomes vertical. The lobster-back now has five full segments of 15 degrees and one half-segment of $7\frac{1}{2}$ degrees. It will therefore be necessary to add another half-segment of $7\frac{1}{2}$ degrees at the other end in order to make up the bend to a full right angle. A bend such as this is shown at Fig. 191. If the new base BC be regarded as circular, then the problem becomes one of development from a right cone. The right cone is shown at BAC, and is divided into five full segments of 15 degrees, a half-segment at the bottom and another at the top.

This illustration, Fig. 191, also shows the effect of reversing the segments to produce the tapered bend. It will be seen that the centreline of this bend cannot lie on a quarter-circle, since the points b,c,d,e,f,g at which the central axis cuts the joint line, fall farther from the true quadrant curve at each reversal. To obtain the spacings a',b',c',d',e',f',g',h' up the centreline of the cone, describe a

quarter-circle of approximately the radius required, and divide it
into six equal parts. The first spacing a,b' up from the base should
be one half of these divisions, then five full spacings, and another
half at the top. Through the first point b' up from the base, draw
a line at $7\frac{1}{2}$ degrees to the base BC. Through the second point c',
draw a line at $7\frac{1}{2}$ degrees to the base, but inclined the opposite way,
thus making an angle of 15 degrees with the first line. Through points

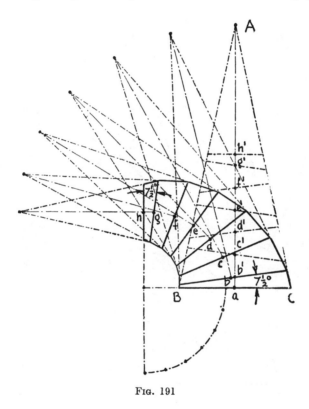

d',e',f',g', draw lines alternately at $7\frac{1}{2}$ degrees as shown, and one
horizontally, or parallel with the base, through the top point h'.
The patterns for these segments may then be developed in accordance with ordinary right conic methods, and as these methods have
been fully discussed in many previous sections, the pattern developments for this problem are left as an additional exercise.

TAPERED LOBSTER-BACK BY COMMON CENTRAL SPHERES

Following up the problem of the tapered lobster-back bend, the method of the common central sphere may be applied to obtain the

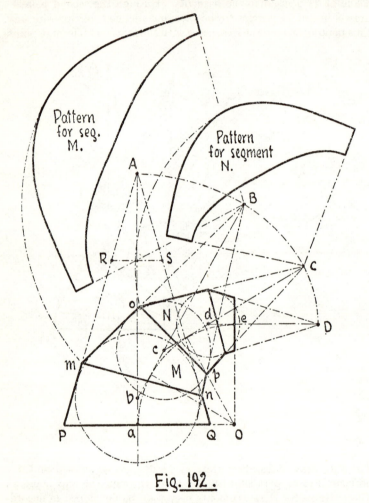

Fig. 192.

segments as portions of right cones, and, in addition, the bend may be set out to predetermined diameters and centre radius. Fig. 192 shows a four-segment bend set out on this principle, together with the patterns for the two middle segments M and N. Fig. 193 illustrates the preliminary steps in setting out the elevation of this bend. First, from centre O, describe a quarter-circle a,e. Divide the right angle aOe into three equal angles of 30 degrees each, and at the points where the division lines cut the quarter-circle draw tangents and produce them to form the centrelines of the four cones. These tangent centrelines are shown at aA, bB, cC, dD. Next, mark off the

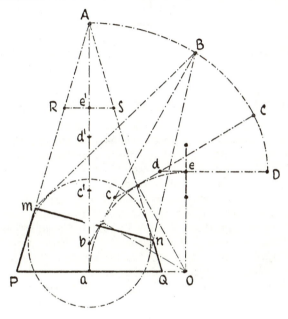

Fig. 193.

lengths of the tangents ab, bc, cd, de up the vertical centreline aA. Through a and e' draw PQ and RS respectively at right angles to aA, and equal in length to the diameters required at each end of the bend. Now draw PRA and QSA, which represent the sides of the first cone. With b as centre, draw the inscribed circle, just touching PA and QA, to represent the first central sphere. Also, with b as centre, describe the arc AB, which locates B, the apex of the second cone. From the apex B draw the sides of the cone just to touch the central sphere. Then where these sides cut the corresponding sides of the first cone, as at m and n, points will be obtained through which to draw the first joint line m,n. The second central sphere may now be drawn, using c as centre, with its circumference just touching the sides of the second cone. See Fig. 192. The apex C, of the third cone may also be located by describing the arc BC from the centre c. Draw the sides of the third cone just touching the second central sphere, and then, between the points where these sides cut the sides of the second cone, put in the second joint line o,p. The third central sphere may now be drawn, using d as the centre, and the processes above repeated in order to complete the fourth cone.

The drafting of the pattern for each segment is straightforward right conic development and has been fully described in an earlier course.

LOBSTER-BACK DEVELOPED BY TRIANGULATION

Patterns for tapered lobster-back bends which have to be made to specified dimensions are best developed by triangulation. Each segment may be regarded as a transforming piece between two circles. In this way almost any combination of segments may be made to make up a tapered bend or a tapered swan-neck. The pattern for each segment has to be developed separately, and, as a problem of pattern-drafting, should not, at this stage, present any real difficulties. Fig. 194 represents a ventilator head, which is composed of six segments. This ventilator head, without the curved rim, is shown again in Fig. 195, but is turned on its side for the purpose of simplifying the problem of drafting the pattern.

Immediately below the base line draw a half-plan of the bottom segment only, by first describing a semicircle on the base 1,13, and dividing it into six equal parts. On the top edge of the segment 2′,14′ also describe a semicircle, divide it into six equal parts and project the points perpendicularly back to the line 2′,14′. From the points on the line 2′,14′, drop vertical lines into the plan, and from the base line mark off the corresponding widths of the top semicircle to obtain points through which to draw the semi-ellipse 2,14. Next, triangulate the plan by numbering and joining the points 1,2,3,4, . . . 11,12,13,14, as shown in the diagram at Fig. 195.

To develop the pattern for the bottom segment of Fig. 196, take the true length line 1,2′ direct from the elevation and mark of 1′,2″ in the pattern. Next, take the plan length 2,3, and mark this off along the base line from the foot of the vertical height below point 2′. Take the true length diagonal up to point 2′ and from point 2″ in the pattern swing an arc through point 3″.

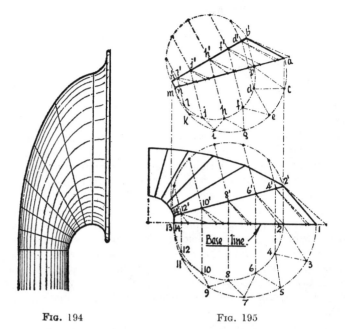

Fig. 194

Fig. 195

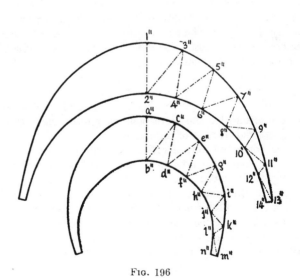

Fig. 196

It will be noticed in this problem that a separate vertical height line, on to which all points are projected, is not given. Instead of this, the vertical height of each point from the base line is used in its primary position. This, of course, is not essential, and the usual vertical height line may be erected if preferred.

Now take the true distance 1,3 direct from the plan, and from point 1″ in the pattern describe an arc cutting the previous arc in point 3″. For the second triangle take the plan length 3,4, and, from the point on the base line vertically below point 4′ in the elevation, mark off this distance at right angles to the vertical height. Take the true length diagonal up to point 4′, and from point 3″ in the pattern swing an arc through point 4″. Next take one of the equal divisions from the top semicircle, and from point 2″ in the pattern describe an arc cutting the previous arc in point 4″. Repeat this process with all the remaining plan lengths between the top and bottom edges, obtaining the true length diagonal for the zigzag line between the curves in the pattern. For the true spacings between the points along the curves in the pattern the corresponding distances should be taken from the semicircle on the base line and the other on the top edge.

The second segment in Fig. 195 is shown redrawn immediately above its original position in the elevation. This is done purely for the sake of clearness in following the development of the pattern. In general the procedure is precisely the same as for the bottom segment just described. The base of the second segment, m,a, should be considered as horizontal, thereby forming a base line similar to that in the bottom segment. The elliptical plan of the top edge n',b' should therefore be obtained by projection at right angles to the base line m,a. For the development of the pattern, the directions given for the previous segment should be carefully followed, substituting the letters for the corresponding figures, since the points have been lettered instead of numbered.

From the foregoing considerations of the various methods of setting out tapered lobster-back bends, it will be evident that the conditions given and the accuracy of the requirements must influence the choice of method used. The bend obtained by the oblique cone method is well formed and of pleasing appearance. That obtained by means of the common central sphere gives a reliable bend on a quarter-circle centreline, while the method of triangulation may be used for any design of tapered lobster-back.

CHAPTER 14

UNUSUAL PROBLEMS

THE value of a clear and versatile knowledge of pattern-developing cannot be over-estimated. It sometimes happens that a working drawing contains a problem of unusual design, or one of a general type with some slight modification which involves a deeper understanding of the geometrical principles of pattern-drafting. Such occasions discover the versatility or the limitations of the craftsman. In the latter case, almost instinctively, suggestions are put forward to simplify the work, often at the expense of efficiency. Discussions follow, and valuable time is expended in determining alterations which mean nothing more than a simplification of the geometry to bring it into line with the vision of the craftsman. Loss of time in this way is often much greater than would be the case if the problem were dealt with promptly to its original design.

The oblique conical hopper shown in Fig. 197 is an example which illustrates this point. The hopper above, at (*a*), is a straightforward oblique conic frustum fitting on a square corner. In the right-hand view the centreline is vertical. By these conditions the pattern can be developed from the left-hand elevation only, an example of which has already been fully described in an earlier course. Now, referring to the figure below, at (*b*), the only difference in the conditions is that the apex, *A*, of the cone in the right-hand view has been moved 2 ft. 7 in. to the right. This may be essential to the required conditions, but it entirely alters the problem from the standpoint of pattern-development. It is still an oblique cone, but the apex is off-centre in both views, and neither view can be used to develop the pattern in the same simple way as in the case of the elevation above at (*a*). However, the solution to the type of problem presented at (*b*) may be arrived at in several ways, all of which are based on the same fundamental principles. The examples given in Figs. 198 and 199 illustrate two different methods of solving for the patterns.

AN OFF-CENTRE OBLIQUE CONICAL BREECHES PIECE

Fig. 198 (*b*) represents a two-way branch piece composed of the frustums of two oblique cones. The only difference between this problem and that given in Fig. 115 in the Third Course is that the branch centres do not lie on the horizontal centreline of the base circle. Nevertheless, the branch piece is symmetrical about the vertical centreline, which makes the two branches similar. The method of developing the pattern illustrated in this figure, at (a), is the same as that adopted for the solution shown in Fig. 115, but the fact that the apex in the plan is off the horizontal centreline,

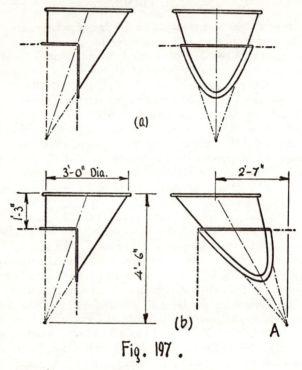

Fig. 197.

the present problem needs more care and insight to press the drafting of the pattern to a successful conclusion.

Assuming that the plan and elevation of one limb have been set down to the required conditions, as shown in the main figure at (a), Fig. 198, the full circular base should be divided into twelve equal parts, as at 1,2,3, . . . 10,11,12. Join these points to the apex A. From the apex A, which occupies a position on the base line produced, swing these plan lengths round to the base line of the elevation and from the base line join the points to the apex A′ in the elevation. These are the TRUE LENGTH lines, and it will be seen that some of them fall outside the elevation of the branch. From the

apex A' in the elevation swing the true length lines into the pattern
Deciding now on the position of the seam, which in this illustration

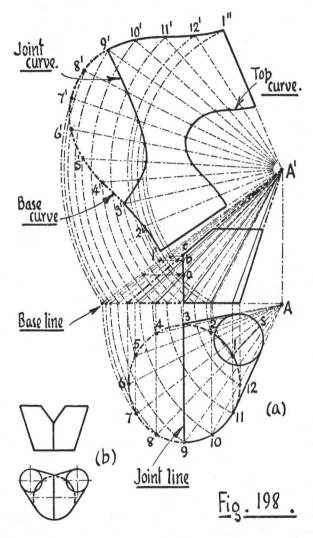

Joint curve.

Top curve.

Base curve

Base line

Joint line

(a)

(b)

Fig. 198.

is taken from 1 to S, as shown in the plan, take one of the equal
divisions from the circular base, and beginning on the arc in the
pattern obtained from point 1, mark off the distances round the base

curve, stepping over from one arc to the next one corresponding to the next point on the circle. Thus, points 2″,3″,4″ fall on the arcs derived from points 2,3,4, and so on. Draw the full base curve through these points. Now join the points on the base curve to the apex A' in the elevation. Next, using the apex A' as centre, swing arcs into the pattern from the points in the elevation where the true length lines cross the top edge of the frustum. Where these arcs, not shown, cross the corresponding radial lines in the pattern, points will be afforded through which to draw the top curve in the pattern.

It now remains to determine that part of the curve in the pattern which represents the joint curve between the two limbs. Project the points 1,2,3, . . . 10,11,12, on the circle in the plan, vertically upwards to the base line in the elevation, and from the points obtained on the base line draw lines to the apex A'. These are the ELEVATION LINES, and are shown as full lines in the diagram at Fig. 198. Where the elevation lines from points 4 and 8, 5 and 7, and 6, cross the joint line in the elevation, draw horizontal lines to meet the corresponding true length lines. Thus, from point 8 the line drawn vertically upwards passes through point 4, giving a common point on the base line. The elevation line from this point passes through point a on the joint line. The horizontal line drawn from point a meets the true length line from point 4, and also, farther along, the true length line from point 8. Now swing these new points into the pattern to meet the corresponding radial lines from points 4′ and 8′. This, repeated from points 5 and 7, and also from point 6, should give the required points through which to draw the joint curve in the pattern, as shown in the diagram.

AN OBLIQUE CONICAL HOPPER

The pattern for the oblique conical hopper illustrated in Fig. 197 (b) is shown developed in Fig. 199. The plan, it will be seen, is projected below the left-hand elevation. The right-hand elevation then becomes unnecessary for the solution of the problem. This problem may be solved in the same way and by the same method as that of the previous example, and may be done by placing the elevation below the plan and swinging the plan lengths round from the apex A to the inverted base of the cone. The procedure would then be similar to that of the problem just described. However, for the sake of variety and versatility, the alternative mode of solution illustrated in Fig. 199 is also given. In the elevation it will be observed that the corner P' occurs on the centreline C',A' of the cone. In setting out the plan, first describe the circle representing the base of the cone, which, since the cone is inverted, is at the top. Next locate the apex A at 2 ft. 7 in. below the centre of the circle, and, of course, vertically below the apex in the elevation. Now, as the plane cutting the cone at $P'Q'$ is horizontal, or parallel to the base, the plan of $P'Q'$ will be a semicircle, and its centre will lie at P on the centreline OA of the cone vertically below P'. Its radius PQ will be

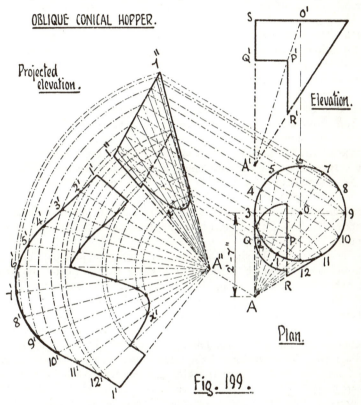

OBLIQUE CONICAL HOPPER.

Projected elevation.

Elevation.

Plan.

Fig. 199.

equal to $P'Q'$. The vertical plane at $P'R'$ will be represented in the plan by a straight line through P as shown in the diagram.

The next step is to obtain the elevation projected at right angles to the centreline OA in the plan. The general principles of projection should not at this stage require detailed explanation, so it is assumed that the projected elevation $1'',7'',A''$ of the full cone will be readily obtained as in the diagram. The semicircle PQ in the plan will appear in the projected elevation as the straight line parallel to $1'',7''$ at a distance from it equal to $Q'S$ in the elevation. To obtain the curved part of the corner cut in the projected elevation, first divide the circle in the plan into twelve equal parts, as shown numbered from 1 to 12. Join these points to the apex A. Next project the points on the circle to the top edge $1'',7''$, in the projected elevation, and join the points obtained on $1'',7''$, to the apex A''. These are ELEVATION LINES and are shown as full lines in the diagram. Now, in the plan, the lines from points 6,7,8,9,10,11,12 all cross the straight line PR. The points where these lines cross PR should be projected into the new elevation to meet the corresponding elevation lines from $1''$, $7''$ to A''. Points will thus be found through which to draw the curve which represents the vertical plane corresponding to $P'R'$.

To develop the pattern, from the apex A swing round the plan lengths $A8$, $A9$, $A10$, $A11$, $A12$, to the centreline $A7$, and project them from there to the top edge $1'',7''$, in the projected elevation. From the points on $1'',7''$, draw lines to the apex A''. These are TRUE LENGTH LINES, and are shown chain-dotted in the diagram. From the apex A'', swing the ture length lines into the pattern. Take one of the equal spacings from the circle, and, beginning on the inside arc from point $1''$, mark off the distances $1',2',3'$, . . . $11',12',1'$, stepping over from one line to the next. Join all these points to the apex A''. The seam, it will be noticed, is located at point 1 in the plan.

To determine the shape of the inner curve, the first part may be found by swinging into the pattern those points where the true length lines cross the straight line representing the semicircle in the projected view. It will be seen in the plan that this part of the curve lies on the semicircle as far round as the line joining point 6 to the apex A. Thus, the first part of the inner curve in the pattern will be obtained as far as the line joining point $6'$ to A'', as shown in the diagram. The next part of the inner curve may be found by projecting the points, where the ELEVATION LINES cross the curve in the projected view, parallel to the base of the cone to meet the corresponding TRUE LENGTH LINES. These points should then be swung into the pattern to meet the appropriate radial lines. Thus, to follow up the process on one only of these points, take the example from point 11 in the plan. The elevation line derived from this point crosses the curve at x. From this point a short line is drawn parallel to the base $1'',7''$ to meet the corresponding true length line derived from point 11. This point on the true length line is then swung into the pattern to meet the radial line in x' between $11'$ and A''. This process, carefully repeated with the elevation lines and true length lines from points 7,8,9,10, and 12 will give the required points in the pattern through which to draw the second part of the inner curve. The remaining short bit between the radial lines $A'',12'$ and $A'',1'$ is a replica of the curve between the corresponding two lines $A'',2'$ and $A'',1'$ on the other side of the pattern.

COMPLEX DOUBLE TRANSFORMER

Many times attention has been drawn to the fact that it is the unusual problem which needs most careful attention in drafting the pattern. An unusual problem is not always an intricate one. Often

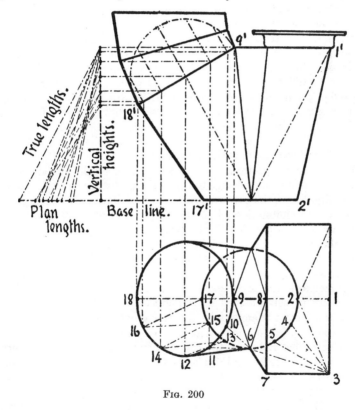

Fig. 200

it is just a familiar type turned round or upside down, or is a combination of elementary parts which only needs to be analysed in order to reveal its simplicity.

The problem shown in Fig. 200 is of a feed connection to a

centrifugal machine. The rectangular side forms the inlet, which is fed from a chute or conveyer. The circular side forms an air connection which assists in the feed of the material through the circular base to the machine. This combination may be developed and made in one piece of metal if desired, although it would be better and easier to make it in two pieces with the seams at 1,2 and 17,18. The rectangular top is connected or joined to one half of the base, while the circular top is joined to the other half. The rectangular side and the circular side are themselves joined together by the triangular pieces as shown on one side at 6,8,9. The metal is bent along the lines 6,8 and 6,9, but there is a joint along the top edge 8,9 between the two triangular pieces.

The surface is triangulated for development as shown in the figure from 1 to 18. The first part from 1 to 7 is similar to a portion of an ordinary tallboy transformer; then come two flat triangles, 6,7,8 and 6,8,9; followed by the transforming portion between the semicircle 9–18, and the quarter-circle 6–17. It will be noticed that the semicircle 9–18 in the plan is seen as a semi-ellipse. This, of course, is obtained by dropping the semicircular edge 9′–18′ from the elevation into the plan in accordance with standard rules, which should be easily followed from the illustration.

To develop the pattern, first erect a vertical height line and project all the points on the edge 9′–18′ horizontally across to it. Then, for the first line in the pattern, take the true distance 1′2′, direct from the elevation and mark off 1″,2″, in any convenient position to begin the pattern, as in Fig. 201. Next, take the plan length 2,3, and mark it off along the base line at right angles to the vertical height. Take the true length diagonal up to the top, and from point 2″ in the pattern swing an arc through point 3″. Now take the true length 1,3, direct from the plan, and from point 1″ in the pattern describe an arc cutting the previous arc in point 3″.

For the second triangle take the plan length 3,4, and mark it off along the base line at right angles to the vertical height. Take the true length diagonal up to the top, and from point 3″ in the pattern swing an arc through point 4″. Next take the true distance 2,4, direct from the plan, and from point 2″ in the pattern describe an arc cutting the previous arc in point 4″.

For the third and fourth triangles repeat this process with plan lengths 3,5 and 3,6, and also with the true distances 4,5 and 5,6, direct from the plan. For triangles five, six, and seven, again repeat the process with plan lengths 6,7; 6,8; 6,9. The corresponding true distances taken direct from the plan are this time 3,7; 7,8; and 8,9.

In the next three triangles, still radiating from point 6, the plan lengths 6,10; 6,11; and 6,12 should be triangulated against the respective vertical heights, but it will be observed that the vertical heights now begin to shorten as the points 10,11,12 occur down the inclined edge 9',18' in the elevation. The true distances between 9 and 10; 10 and 11; 11 and 12 should not this time be taken from the plan, but from the spacings round the semicircle on the edge 9',18'

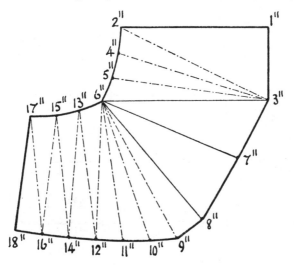

Fig. 201.

in the elevation. The remainder of the pattern is straightforward triangulation, and should be easily completed by following the method already described.

PITFALLS IN TRIANGULATION

Although the method of triangulation may be readily applied to the solution of problems which cannot be solved by the other two methods, and in some ways may appear to be the simple and ideal mode of pattern-drafting, there are pitfalls which mar the success of its application unless care be exercised in the preliminary process of dividing the surface of the object into triangles. One of the essential conditions is that the straight lines forming the triangles shall lie on the surface of the metal, or as near that condition as possible.

Any line placed crosswise on a curved surface must introduce error into the pattern. The extent of the error depends on the difference between the length of the curved line and that of the straight line between its extemities.

Fig. 203 shows a connecting duct between a rectangle on a vertical cylindrical surface and a circle meeting the lobster-back segments.

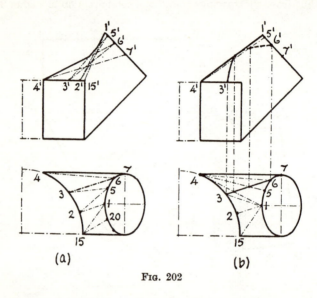

FIG. 202

The elevation in this figure is a typical representation of the connecting piece as it would be given on a working drawing. In applying triangulation to this problem it is probable that many exponents would advocate dividing the surface as shown in the plan at Fig. 202 (a). The lines forming this arrangement are shown again in the elevation above, and it will be seen that not one of them reaches the straight top line shown in the elevation at Fig. 203. This indicates that the arrangement of triangulation shown in the plan at Fig. 202 (a) does not meet the requirements of the original drawing, and that the surface triangulated in this way would produce an elevation with a curved or broken top line as shown at (a), Fig. 202. The metal, in this case, would require beating or distorting to make it conform to the required dimensions. Now, referring to the diagram at (b), Fig. 202, the top point 1' in the elevation occurs at point 1 in the plan, and similarly point 4' occurs at point 4. Therefore, the straight

line 4′,1′ in the elevation must occur at 4,1 in the plan, and this
line represents the surface of the metal between those two points.
Now, from point 4 it is an easy matter to produce a conical surface

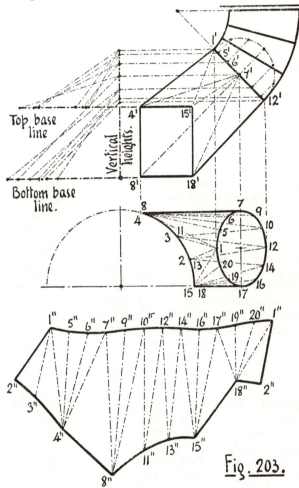

Fig. 203.

with lines radiating to points 1,5,6,7, and a similar conical surface
from the opposite point 1, with lines radiating to points 4,3,2,15,
as seen in Fig. 202 (b). If, in these circumstances, a line were considered as passing from point 3 to point 6, which corresponds to one
of the lines in the plan at (a), Fig. 202, it can be shown, as in the

elevation at (b), Fig. 202, that this line is considerably curved; a condition which should be rigorously avoided. The failure of the arrangement of the triangulation shown at (a), Fig. 202, is due to the two edges 4,3,2,15 and 7,6,5,1,20, curving in opposite directions. The same arrangement is quite satisfactory on the two bottom edges, 8,11,13,18 and 9,10,12,14,16, since these curve in similar directions, as will be seen in Fig. 203. This illustrates a common pitfall in the application of triangulation. It does not affect the principles of pattern-development by that method, but represents the possibilities of error arising out of the faulty division of the surface into triangles.

SPECIAL TYPE HOOD CONNECTION

The pattern for the hood connection shown in Fig. 203 has to be developed in full, as it is not symmetrical about any axis. Assuming that the plan and elevation have been set down to required dimensions, and that the circular edge 1',12 in the elevation appears in the plan as the ellipse 1,12, then, beginning at the position of the seam 1,2, divide the surface into triangles in the manner shown in the diagram. Erect a vertical height line in the elevation and project all the points on the edge 1',12' horizontally to meet it. Two base lines should also be taken out level with 4',15' and 8',18'.

For the pattern, take the plan length 1,2, and mark it off along the top base line at right angles to the vertical height. Take the true length diagonal up to the top point level with 1', and in any convenient position mark off 1",2" in the pattern. Next take the plan length 1,3, and mark it off along the top base line at right angles to the vertical height. Take the true length diagonal up to the top point level with 1', and from point 1" in the pattern swing an arc through point 3". Now take the true distance 2,3 direct from the plan and from point 2" in the pattern describe an arc cutting the previous arc in point 3". For the second triangle, repeat this process with plan length 1,4, and the true distance 3,4 from the plan. For the third, fourth and fifth triangles, repeat the process with plan lengths 4,5; 4,6; 4,7; but the true distances between points 1 and 5, 5 and 6, 6 and 7, must this time be taken from the semicircle on 1',12', in the elevation. The next triangle, number six, lies in a vertical plane, and in the plan is represented by the straight line 4,7,8 in which point 4 and point 8 occupy the same position. In the elevation this triangle is shown at 4',7',8', and, because its position in the plan is horizontal, the elevation view represents its true size. Therefore, from the elevation, take the true distance 7',8', and from point 7" in the pattern swing an arc through point 8". Next take the true distance 4′,8′, direct from the elevation, and from point 4″ in the pattern describe an arc cutting the previous arc in point 8″. For the next two triangles, the plan lengths 8,9 and 8,10 should be triangulated against the full vertical height from the base line, and the true distances between 7–9 and 9–10, should be taken from the

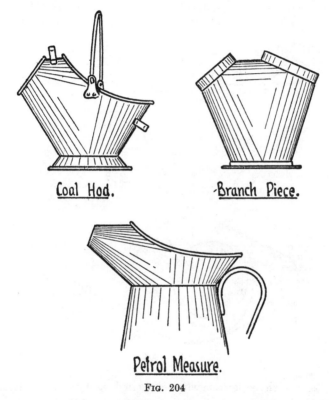

Coal Hod.   Branch Piece.

Petrol Measure.

FIG. 204

semicircle on 1′,12′. From this point the remainder of the pattern should present no difficulty, provided care be taken to triangulate the plan lengths against their respective vertical heights, and that only true distances are taken for inclusion in the pattern.

COAL HOD AND PETROL MEASURE TOP

The form of the coal hod shown in Fig. 205 represents a type of problem which is adaptable to several uses. Some of the purposes to which it applies are shown at Fig. 204. The top for the petrol measure is much used as a safety pouring device, while the branch piece is sometimes favoured in preference to other designs.

Development of the pattern by triangulation is not difficult, provided that the usual care is taken to place each plan length against its appropriate vertical height. This remark applies chiefly to the top portion, seen in the elevation between the points 7′,9′,21′,20′, Fig. 205. The large hole at the top, 1′ to 21′, is circular in the plan,

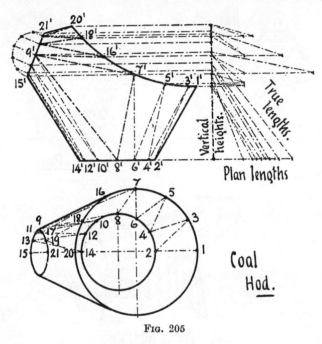

FIG. 205

but its true shape is somewhat elongated, and the true distances between the points on its circumference must not be taken direct from the plan, but obtained by triangulation. The smaller hole at the top, 15′,21′, is circular in its actual form, and therefore appears as the ellipse 15,21 in the plan. As the plan is symmetrical about its horizontal centreline, only one half of the surface is triangulated for the pattern. To divide the surface into triangles, join the quadrant 2,8, in the bottom to the quadrant 1,7, in the top by the zigzag line 1,2,3,4,5,6,7,8. This line in the elevation is represented by the points 1′,2′,3′,4′,5′,6′,7′,8′. Then comes the large triangle, seen best in the elevation at 7′,8′,9′, and in the plan at 7,8,9. The next part of the surface, again best seen in the elevation, is divided into triangles by the zigzag line from 8',9' to 14',15', and in the plan as seen at 8,9,10,11,12,13,14,15. The remaining part of the surface to be divided into triangles occurs in the elevation from 7',9' to 20',21', and in the plan between the points 7,9,16,17,18,19,20,21. In the plan, this last part to be triangulated somewhat overlaps the previous part, but with care in numbering the points the solution should not be troublesome.

To develop the pattern, erect a vertical height line, and project all the points on the two top edges 1',20' and 15',21' horizontally

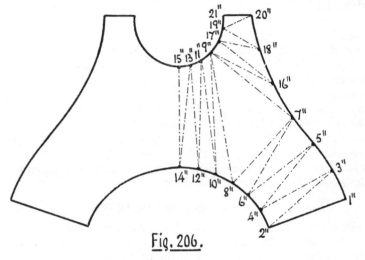

Fig. 206.

on to it. For the first triangle take the true distance 1',2' direct from the elevation and mark off 1",2" in any convenient position to begin the pattern. (See Fig. 206.) Next take the plan length 2,3, and mark it off along the base line at right angles to the vertical height. Take the true length diagonal up to the point level with 3', and from point 2" in the pattern swing an arc through point 3". Now take the true distance 1,3 direct from the plan, and from point 1" in the pattern describe an arc cutting the previous arc in point 3".

For the second triangle, take the plan length 3,4 and mark it off at right angles to the vertical height. Take the true length diagonal up to the point level with 3' and from point 3" in the pattern swing an arc through point 4". Next take the true distance 2,4, direct from the plan, and from point 2" in the pattern describe an arc cutting the previous arc in point 4".

For the third triangle, take the plan length 4,5, and mark it off at right angles to the vertical height. Take the true length diagonal up to the point level with 5', and from point 4" in the pattern swing an arc through point 5". Now, for the true distance between points 3 and 5, as the top edge now begins to rise, the plan length should be triangulated against the vertical height by marking off the plan length 3,5 along the line level with point 3', and the true length diagonal taken up to the point level with 5'. Then from point 3" in the pattern describe an arc cutting the previous arc in point 5".

For the fourth, fifth, and sixth triangles, repeat this process with plan lengths 5,6; 6,7; 7,8; taking the true distances 4,6 and 6,8 for the bottom edge direct from the plan. The true distance for the top edge, between points 5 and 7, should be obtained by triangulation as in the case of the distance between points 3 and 5.

The next triangle 7",8",9" is obtained by triangulating the plan lengths 7,9 and 8,9 against their respective vertical heights, and marking them off in the pattern as shown in the diagram. From this point the system of triangulation takes two directions; one from the line 8,9 to the line 14,15; and the other from the line 7,9 to the top line 20,21. With care in following the method just described, the remainder of the pattern should not be difficult to complete. The true distances for the small circular hole should be taken from the semicircle 15',21' in the elevation.

CONICAL HOPPER CORNER

It is sometimes required that the corners of a tapered hopper should take a radius in the form shown at (a) in Fig. 207. The flat sides of the hopper, produced, taper to a central point or apex. Without a radius the corners would be as shown at (b), Fig. 207. The radius at the top at (a) is larger than that at the bottom, and the two respective centres are shown at O and o. O is the centre of the quadrant O,b,c, and o is the centre of the quadrant o,d,e. A line drawn through O,o and two more through b,d and c,e should all meet at a. It should be clear from the diagram that the point a is the apex of a quarter of an oblique cone, as seen at a,d,b,c,e. The development of the pattern for this corner is easy and straightforward when dealt with by triangulation, but as an oblique conic problem, which is not a difficult one, it may offer some degree of interest to the student of sheet metal geometry. The solution on oblique conic lines is given in Fig. 208.

In the plan of the corner, Fig. 208, the centreline of the cone is shown at a,O,f. The quadrant O,b,c forms the base and is divided into three equal parts, as at b,g; g,h; and h,c. The elevation a',O',f' is projected at right angles to the plan centreline a,O,f, and the height HT represents the vertical depth of the hopper.

To develop the pattern, with a in the plan as centre, swing the points b,g,h,c round to the centreline a,O,f. Project the points obtained on the centreline into the elevation to O',f'. With a' as centre, swing arcs from the points obtained on O',f' into the pattern. Also, from the points on O',f', draw lines to the apex a'. From the

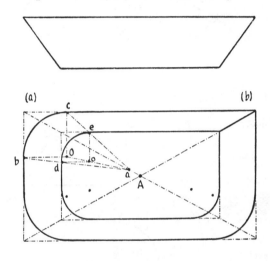

Fi9. 207.

points where these lines cross the bottom base line o', again swing arcs into the pattern. Now take one of the equal distances b,g; g,h; or h,c from the quadrant in the plan, and, beginning on the inside arc from the top, mark off the distances c'',h'',g'',b'', stepping over from one arc to the next, as shown in the diagram. Draw in the top curve. Join the points on the top curve to the apex a', and where these radial lines cross the inside arcs from the bottom, points will be afforded through which to draw the bottom curve.

That part of the pattern shown between the points b'',c'',e'',d'' represents the oblique conical corner. It still remains to obtain the relative position or direction of the straight sides as determined by the lines b'',m'' and c'',n''. In the pattern the lines d'',m'' and e'',n'' represent the true depths of the side plates, and to obtain these take the plan lengths d,m and e,n, and triangulate them against the vertical height HT in the elevation. Take the true length diagonals, and from the points d'' and e'' describe arcs with the respective true

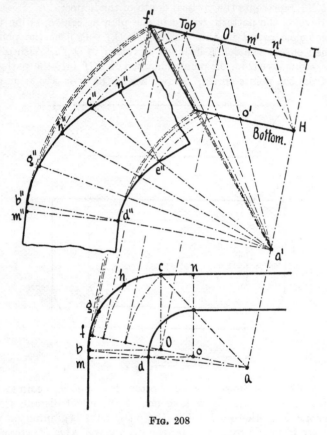

Fig. 208

lengths through m'' and n''. Then draw b'',m'' tangent to the first arc, and c'',n'' tangent to the second arc. The corresponding lines from d'' and e'' may now be drawn parallel to d'',m'' and c'',n''.

FIFTH COURSE

Chapter 15

DOUBLE PROJECTION

When systems of pipe work are designed by the engineer or draughtsman, they are usually set out on drawings in accordance with the principles of orthographic plans and elevations. These drawings are those which find their way to the workshop when the order is placed for construction. The draughtsman rarely takes into consideration the placing of views on the drawing which will help the craftsman to draft his patterns. In consequence, of this, many simple problems of pattern-development involve difficult problems of projection before the pattern can be drafted. It is true that some gifted craftsmen can apply rule-of-thumb methods with reasonable success, but when accuracy is an important condition, methods of chance are not always satisfactory nor profitable. Against this, draughtsmen themselves do not always draw their views correctly, which is not helpful when work has to be "made to drawing." Therefore the problems dealt with in this course are not set down in their simplest form, ready for development, but are presented as they would be given on standard working drawings.

The example shown in Fig. 209 is a simple section of duct work showing an offset in a line of pipe with a branch in the middle. The plan is almost exactly the same as the elevation, yet there is nothing peculiar about the conditions given in this problem. They are quite commonplace, and the views are correctly presented. There are two points of interest arising out of this drawing which would have to be settled satisfactorily in the course of construction. One is the actual angle required for the elbows, and the other is the development of the branch pipe at the intersection.

The correct representation of the joint lines in the two views at Fig. 209 is a problem of draughtsmanship, and assuming that these lines were correctly drawn, the pattern for the branch piece could be developed, or "unrolled," direct from either the plan or the elevation. However, the stages of projection required to produce true lines of intersection are rarely considered necessary by the draughtsman, and the joint lines are more often drawn

277

approximately. Nevertheless, to develop the pattern correctly, a true line of intersection must first be found.

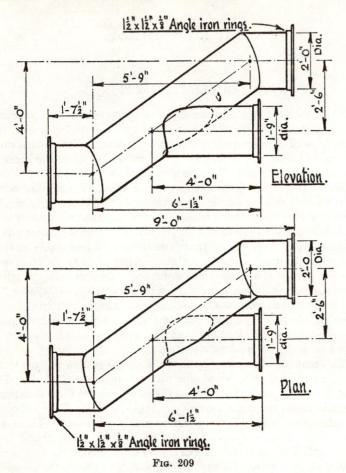

Fig. 209

DOUBLE PROJECTION APPLIED TO PIPE INTERSECTION

To determine the line of intersection, sometimes two or even three projections are necessary. In problems of this kind the object is to obtain a view of the pipe and branch as they would be seen lying flat on the paper. It will be observed that neither the plan nor the elevation in Fig. 209 fulfils this condition. Referring now to Fig. 210, the first stage is to project a side elevation in the direction of

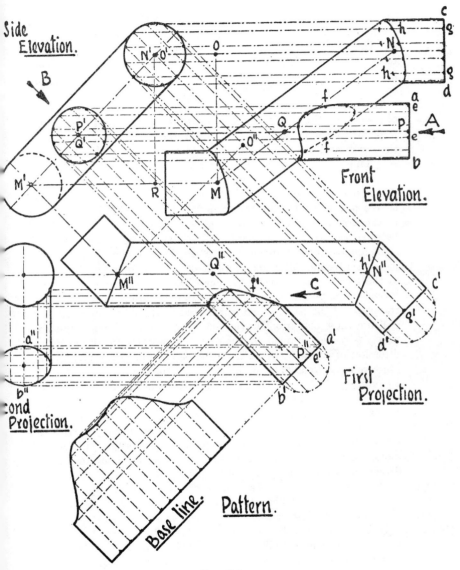

Side Elevation.

B

Front Elevation.

First Projection.

Second Projection.

Base line. Pattern.

Fig. 210

arrow A, in which R,M' is the off-centre distance taken from the plan. The second stage is to project a view in the direction of arrow B. This projected view is the one in which the pipe and branch lie flat on the paper. The third step is to obtain a second projected view in the direction of arrow C, from which the joint line is then found by projecting back to the first projection.

The important principle governing these projections is the transfer of heights or distances from one view to its complement in the third position. Thus, in the Front Elevation, the distance O,N represents the horizontal distance between the two intersection points M and N. This distance, O,N, when turned up into the side elevation, becomes the vertical distance, at right angles to the plane of the paper, between the points M' and N', and in this view the point N' covers point O'. In the next step the distance O,N is transferred to the first projection and becomes the projected distance O'',N'' between the intersection points M'',N''. Since this projection is at right angles to the centreline M',N' in the side elevation, the projected centreline M'',N'' is its TRUE LENGTH, and therefore lies flat in the plane of the paper. Similarly, as the centreline P,Q of the branch piece becomes vertical in the side elevation, it will lie flat in the plane of the paper when transferred to the first projection.

From these conditions in the first projection, the true joint line may easily be obtained by making yet another projection in the direction of arrow C, in which the cross-section of the pipe M'',N'' appears as a circle. The circular end a',b' of the branch appears as the ellipse a'',b''. It is assumed that the reader is acquainted with the elementary principles of projection, and it should suffice therefore to indicate that the ellipse is projected from a',b' in the first projection. The points obtained on the ellipse are projected to the circle, and from the circle back to meet the lines projected from a',b' along the surface of the branch pipe. Thus the true joint line may be drawn in through the points of intersection.

The pattern may now be "unrolled" in the usual way, as shown in the illustration at Fig. 210. The actual angle of the elbows is also obtained in the first projection, as is shown at each end of the pipe.

In order to obtain the true line of intersection in the front elevation, and also the true elliptical form of the joint lines at the elbows, it will be necessary to project back from the first projection to the side elevation, and from there to the front elevation. Thus, the points on a',b' obtained from the semicircle thereon should be projected back to the circle representing the branch pipe in the side elevation. The points obtained on this circle should then be projected horizontally to a,b in the front elevation. Now, from the branch pipe in the first projection take the distances from the edge a',b' to the joint line in that view, and mark these off along the corresponding lines from a,b in the front elevation. For example, take the distance e',f' from the first projection and, following the line back to the circle in the side elevation, and from there to the base a,b of the branch in the front elevation, mark off e,f on that line which corresponds to e',f'. It will be seen that the line from e',f' cuts the circle in the side elevation in two points which correspond to similar lines on opposite sides of the branch pipe. Therefore, the distance e',f' should be marked off along both of the corresponding lines in the front elevation, as shown in the illustration. This process, repeated with the other distances from the branch pipe in the first projection, should afford sufficient points through which to draw the true intersection curve in the front elevation.

Now, to obtain the elliptical joint line at N in the front elevation, project the points on the semicircle c',d' in the first projection back to the corresponding circle in the side elevation, and from there project the points horizontally to the edge c,d in the front elevation. A similar process of transferring the distances from the first projection to the front elevation will afford means of producing the ellipse. For example, take the distance g',h' from the first projection, and, carefully following the line back through the side elevation to the front elevation, mark off g,h on the two corresponding lines. This process repeated with the other lines on the elbow piece should give sufficient points through which to draw the true elliptical shape of the joint at the elbow in the front elevation. The joint line at the opposite end of the section may be obtained in the same way, or repeated by any other means, since it is exactly the same in form as the one just described.

A SHORTER METHOD

While the above description is somewhat full in the matter of projections, it is realized that most craftsmen will be concerned only with that part which produces the pattern. With this in view, most of the operations can be performed without actually drawing the views, except the branch piece itself, of which the pattern is required. In Fig. 211 the centrelines only are manipulated in order to arrive at the position of the branch pipe. The triangle M,N,O corresponds to triangle M,N,O in Fig. 210, and the distance R,M' to the off-centre position of M'. Then the line N',N'' is drawn at right angles to M',N', and the distance O,N'' is cut off equal to O,N. P,Q, the centreline of the branch piece, is produced to P', and then projected at right angles to M',N' to cut M'',N'' in Q''. The distance P'',Q'' is then cut off equal to P,Q, which is the centreline of the branch in the required projection.

Now, through P'', at right angles to P'',Q'', draw the base of the branch piece and produce it to may the base line in the pattern.

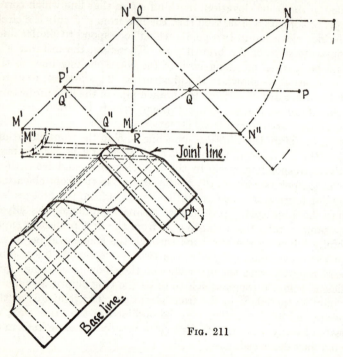

Fig. 211

On the base of the branch describe a semicircle, divide it into six equal parts, and project the points through to the joint line. Next, with M'' as centre, describe the two quadrants, the outer with a radius equal to that of the larger pipe, and the inner with a radius equal to that of the branch pipe. Divide the inner quadrant into three equal parts, and from the points thereon drop lines perpendicular to M'',N'', to cut the outer quadrant. From the points on the outer quadrant draw lines parallel to M',N', to meet the lines projected from the base of the branch pipe. Thus, points will be obtained for drawing in the line of intersection, and the pattern may be "unrolled" in the usual way.

MORE PROJECTIONS

The illustration at Fig. 212 is a further example of pipe work, presented in ordinary plan and elevation, which contains problems of development requiring two or more projections before the patterns can be drafted. The connecting piece to the corner of the rectangular duct forms an elbow at E with the horizontal pipe. The true angle of the elbow is not seen in the plan nor the elevation, but this angle has to be found in the course of solution. The horizontal pipe E,F divides at F into two branches of equal diameter, which turn into the

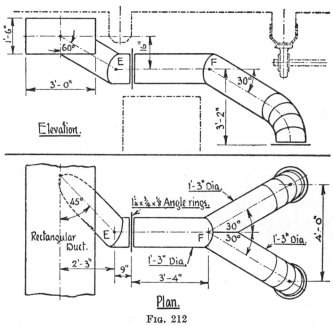

Fig. 212

vertical to meet uptake pipes from two hoods. The hoods are not shown. The horizontal pipe and the two branches from a "Y" piece which, if all three were in one plane, would be a very easy problem of development, but since the two branches dip at an angle of 30 degrees to the horizontal pipe, a double projection is needed to obtain the conditions necessary for developing the pattern.

A DIFFICULT " Y " PIECE

The solution of the above problem is given in Fig. 213. The first projection is equivalent to a side elevation, and the second projection

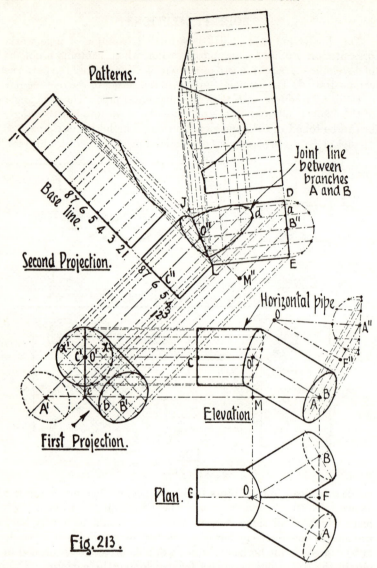

Patterns.

Joint line
between
branches
A and B

Base line.
8 7 6 5 4 3 2 1

Second Projection.

8 7 6 5 4 3 2 1

Horizontal pipe.

First Projection.

Elevation

Plan.

Fig. 213.

produces the conditions from which the patterns are "unrolled." In
the first projection the distance between the points *A'* and *B'* should
be equal to the distance between the points *A* and *B* in the plan.
The points *A* and *B* in the plan may be marked off in any convenient

position along the centrelines of the branches, but both equal in distance from the centre point O. Thus, in the elevation, point B lies behind point A. Now, the object of the second projection is to obtain a view of the horizontal pipe with one of the branches as they would be seen lying flat in the plane of the paper. This is done by taking the second projection in the direction of the arrow at right angles to the centreline in the first projection.

First mark off the position of the base line through C'', and then mark off the distance $C''O''$ equal to C,O from the elevation. Mark off O'',M'' at right angles to C'',O''. Next project point B' into the second projection, and from the point M'', mark off M'',B'' equal to M,B from the elevation. Join O'',B''. Then C'',O'',B'' represents the true angle of the elbow between those two pipes, and the joint line is the straight line J,L, which bisects the angle C'',O'',B''. A straight line D,E through B'' at right angles to O'',B'' may now be drawn to represent the end of the pipe, and the pipe itself drawn in. Describe a semicircle on D,E, divide it into six equal parts and project the points perpendicularly back to D,E. From the points on D,E, project lines into the first projection to obtain the ellipse around the centre B'.

The next step is to obtain the elliptical curve in the second projection which represents the joint line between the two branches A and B. From the points on the ellipse around the centre B', draw lines parallel to the centreline O',B', to cut the circle around the centre O'. From the points on the circle draw lines into the second projection to meet the joint line J,L. From the points on the joint line J,L, project lines on to the edge of the pipe D,E, thus completing the circuit.

Now, some of the lines in the first projection parallel to O',B' cross the joint line between the branches A and B, as, for example, where the line from point b crosses the joint line at c. All the points on the joint line should now be projected into the second projection straight through to meet the corresponding lines on the branch pipe. For instance, the point c is projected through to meet the corresponding line at d. It will be observed that the line b,c in the first projection is the line corresponding to a,d in the second projection. This process repeated with the other points should provide sufficient through which to draw the half-ellipse in the second projection. The pattern for that branch may now be "unrolled" in the usual way as shown in the illustration. The seam in this case is made along the line L,E, instead of the shortest, which somewhat simplifies the pattern-development.

The pattern for the pipe on C'',O'' may also be "unrolled" as shown in the diagram, by projecting the points on the joint line J,L into the pattern for the contour. It will be seen that a smaller semi-ellipse is shown below the joint line J,L. This represents the part of the joint line which occurs between that pipe and the branch A', as it would appear in the second projection, but remembering that the joint line on this pipe is similar for both branches, that semi-ellipse is not essential for the drafting of the pattern. Thus, the half-pattern from 1 to 8 is repeated in the reverse order from 8 to 1'.

However, the semi-ellipse may be readily plotted by locating the points where the lines, projected into the pattern from the joint line J,L, cross the symmetrically opposite lines rising from the first projection. For example, in the first projection the point marked x lies at the same height or level as the point symmetrically opposite marked x'. Therefore, in the second projection, where the point on J,L, which corresponds to point x, is projected into the pattern and crosses the line rising from x', the position of x' will be located on the semi-ellipse. This process applied to the other points on the opposite side of the joint line in the first projection will locate sufficient points through which to draw the semi-ellipse.

In order to obtain the true elliptical shape of the joint line in the elevation, the points on the circle in the first projection should be projected into the elevation, and then, from the second projection, the heights or distances from the base line through C'' to the joint line J,L should be taken and marked off along the respective lines from the base line through C in the elevation. The elliptical curve may then be drawn in.

The elliptical shape of the branch end at A in the elevation may most easily be obtained by setting off the projected edge as shown in the diagram. The distance F'',A'' should be marked off equal to F,A, in the plan, and the pipe edge drawn through A'' at right angles to O,A''. The method of projecting the ellipse back into the elevation should now be evident from the illustration.

THE WORK SHORTENED

The projections dealt with in the above problem have been shown in full, but in cases where the patterns only are required much of the work may be omitted if the processes of projection be clearly understood. The illustration at Fig. 214 shows the problem arranged for the development of the patterns. The centrelines only are used to obtain the second projection from which the patterns are drafted. In the composite plan, Fig. 214, which is equivalent to the first projection in Fig. 213, the centrelines C,O,B represent those from the elevation. The centreline O,B' is similar to that set out in the first projection, in which M,B' is equal to half of the distance between A and B in the plan. The second projection is now made at right angles to O,B', and the distance O,O'' marked off equal in length to C,O. Next, O'',M'' is marked off at right angles to O,O'', and the

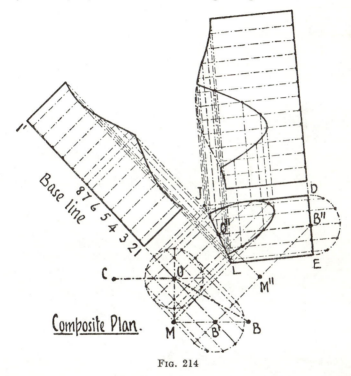

FIG. 214

height M'',B'' made equal to the distance M,B. The line D,E, representing the end of the pipe, is now drawn through B'' at right angles to O'',B''. In the composite plan, instead of drawing the elliptical end of the pipe, a straight line through B' may be drawn at right angles to O,B', and made equal to the diameter of the pipe. Next, on this diameter through B', and also on D,E, draw semicircles, divide each into six equal parts and project the points through to meet, in the one case, the circle around O, and in the other case the joint line J,L.

The procedure now for obtaining the elliptical joint line in the second projection is precisely the same as in the previous example. The patterns may also be unrolled in the same way.

ELBOW CONNECTION TO RECTANGULAR PIPE

In order to effect a solution of the problem of the elbow connection to the rectangular pipe shown in Fig. 212, the object will be to obtain

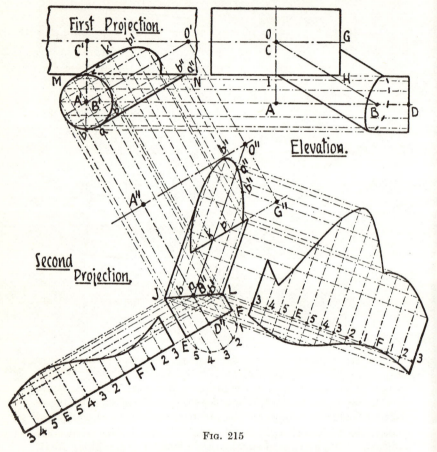

Fig. 215

a view of the elbow as it would be seen lying flat in the plane of the paper. The first projection is taken in the position of a side elevation, see Fig. 215, and the second projection at right angles to the centre-line O',B' of the branch connection. In the first projection the distance O',C' is equal to the corresponding off-centre distance taken from the plan, which is not shown in Fig. 215. Next, to obtain the centrelines in the second projection, mark off A'',O'' in any convenient position parallel to A',O', and from A'' mark off A'',B'' equal to A,B in the elevation. Produce A'',B'' to D'', making B'',D'' equal to B,D in the elevation. Then O'',B'',D'' represents the true angle of the elbow, and J,L the joint line between the pipes.

Next, the elliptical curves in the first and second projections representing the joint lines have to be obtained. To do this, describe a semicircle on the end of the pipe E,F in the second projection, divide it into six equal parts and project the points perpendicularly back to E,F, and on through the joint line J,L to the circle in the first projection. From the points where the lines meet or cut the circle in the first projection, draw lines parallel to the centreline B',O'. From the points where the lines cross the joint line J,L, draw lines parallel to the centreline B'',O''.

The elliptical joint line in the second projection may be regarded as a part of a complete ellipse which would occur if the pipe were entirely cut by the horizontal base M,N of the rectangular duct, as seen in the first projection. Thus, the complete ellipse may be plotted by projecting all the points where the lines on the branch pipe meet M,N into the second projection to meet the corresponding lines on the branch pipe. For example, in the first projection, the lines a,a'' ; b,b'' ; and b',b'' meet the horizontal line M,N, in a'' and b''. Then points a'' and b'' are projected into the second projection to meet the corresponding lines from points a,b and b'. This process repeated with the other points on M,N should afford sufficient points through which to draw a full ellipse in the second projection. However, only that part of the ellipse is required which lies above the line through G''. The distance O'',G'' is marked off equal to O,G half of the width of the rectangular duct in the elevation.

That part of the ellipse which occurs in the first projection is the joint line which lies up the side G,H of the rectangular duct, and may be plotted by projecting the points on the line G'' back from the second projection to the corresponding lines in the first projection. For example, the point p in the second projection corresponds to the point p' in the first projection. Similarly, the point k corresponds to point k'. With a little care the remaining points on that curve may be found by following this reverse projection, as shown in the diagram.

The patterns may be obtained from the second projection by "unrolling" in the usual way. The seam is taken as from B'' to D'' and B'' to k. In view of the many examples already described, a detailed explanation of the pattern-development is not here given, but left as an additional exercise for the reader or student.

AN OFF-SET PIPE CONNECTION

The off-set pipe connection to the grading machine, shown in Fig. 216, is another example which requires a double projection in order

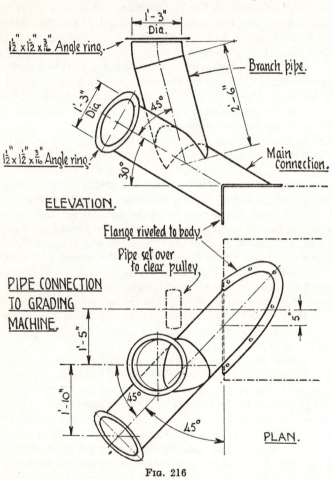

FIG. 216

to arrive at the conditions necessary for "unrolling" the pattern. The main pipe in the plan is shown set over to miss the belt drive of the machine, and this off-set not only adds to the difficulty of

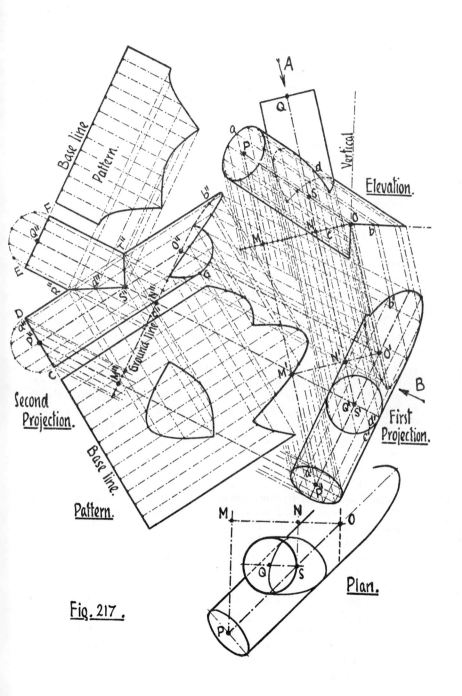

Base line

Pattern.

a

P

Q

A

d

S

Vertical

Elevation.

b″

O″

c″

M

N

c

O

b

G

e″

T″

P″

D

d″

M″

N″

Ground line

S′

b′

O′

Second Projection.

C

P″

M′

N′

M

O

B

G″

S″

d′

c′

First Projection.

Base line.

Pattern.

Fig. 217.

M

N

O

Q

S

P

Plan.

developing the patterns, but complicates the problems of draughtsmanship in setting out the joint lines correctly in the elevation and
plan. The branch pipe forms an alternative feed connection to the
machine.

As in the previous problems of this kind the object is to obtain a
view of the pipes as they would be seen lying flat in the plane of the
paper. The first projection, Fig. 217, is a view looking into the branch
pipe, that is, in the direction of arrow A. In this view the branch
pipe is presented as a plain circle. The second projection is then
taken at right angles to the central axis of the main connecting pipe
in the direction of arrow B, as shown in the diagram at Fig. 217.
The second projection is the view required, in which the two pipes
lie flat in the plane of the paper. This means that the central axes
are seen at their true lengths, and at the true angle in relation to
each other. Since the two pipes are of equal diameter, the line of
intersection is composed of the two straight lines as shown at
R'',S'',T'''.

To obtain the first projection the central axis Q,S of the branch
pipe should be projected to Q',S', which is a single point representing
that axis at right angles to the plane of the paper. From the point
S', mark off S',N' back along the line of projection equal in length
to the distance S,N taken from the plan. Through N' draw M',O'
at right angles to the line of projection. Next, project points P and
O from the elevation into the first projection, the projectors passing
through and locating the points M' and O'. Now mark off M',P'
equal in length to the distance M,P taken from the plan. A line
drawn through P',S',O' represents the central axis of the main pipe
in the first projection. The circle around S' may now be drawn to
represent the branch pipe, and also the lines tangent to the circle
parallel to P',S',O' to represent the main pipe in the first projection.
The ends of the main pipe have yet to be drawn. From the point O in
the elevation, draw M,N,O at right angles to the lines of projection,
thus making it parallel to M',N',O' to serve as a ground line.

The next step is to project the point S' into the second projection,
and mark off S'',Q'' equal to the distance S,Q taken from the elevation. Also, on the same line of projection, mark back S'',N'' equal
to the distance S,N taken from the elevation. Now, through N''
draw M'',N'',O'' at right angles to the line of projection. This may
be regarded as a ground line in the second projection which corresponds to its counterpart, M,N,O, in the elevation. The points P'
and O' may now be projected into the second projection, passing
through and locating the points M'' and O''. Mark off M'',P'' equal

to the distance M,P taken from the elevation. A line drawn through P'',S'',O'' represents the central axis of the main pipe in the second projection, and the line S'',Q'' represents the central axis of the branch pipe. Both of these axes are now in a plane corresponding to that of the paper, and are therefore seen in true length and relation to each other. Through P'', draw C,D at right angles to O'',P'' to represent the end of the main pipe. Similarly, E,F, drawn through Q'' at right angles to S'',Q'', will give the end of the branch pipe. The two pipes may now be drawn in, and also the joint line R'',S'',T''. The shape of the other end of the main pipe at O'' has yet to be obtained.

In order to determine the shape of the end of the main pipe at O'', which fits on the angular corner of the machine body, the process of projection must be in the reverse order. On the end of the pipe C,D describe a semicircle and divide it into six equal parts. Project the points perpendicularly back to C,D, and produce them along the pipe parallel to the central axis P'',O''. Next project the circular end C,D, back into the first projection to obtain the ellipse around P'. The points obtained on this ellipse, which correspond to those on C,D, should now be projected along the pipe parallel to the central axis P',O'. These lines also correspond to those already drawn on the pipe surface in the second projection. The next step is to project the ellipse from the first projection back into the elevation. In this projection it will be observed that the lines from the ellipse do not cross the diametrically opposite points, and this results in twelve projector lines instead of six or seven. In order to locate the points on the ellipse in the elevation, take the perpendicular heights from the ground line in the second projection up to the line C,D, and mark these heights up on the corresponding lines from the ground line M,N,O in the elevation. The points thus located should provide sufficient through which to draw the ellipse representing the end of the main pipe in the elevation. Now, from the points on this ellipse draw lines parallel to the central axis P,O, to meet the angular corner of the machine, as shown in the diagram. The next step is to draw lines from the points on the angular corner to meet the corresponding lines on the pipe surface in the first projection. Thus, in the elevation the point a on the ellipse corresponds to point a' in the first projection, and the line a,b corresponds to the line a',b' in the respective views. Then the line drawn from b into the first projection meets a',b' in point b'. This process repeated from all the other points on the angular corner should provide sufficient points in the first projection through which to draw the contour representing the joint line on the angular corner.

This joint line or contour may now be transferred to the second projection by projecting all the points on the contour in the first projection to the corresponding lines in the second projection. Thus, in the second projection the line a'',b'' corresponds to the line a',b' in the first projection. Then, the line drawn from point b' into the second projection should meet the line a'',b'' in point b''. In this way the contour representing the joint line on the angular corner may be obtained in the second projection.

The patterns for both pipes may now be "unrolled" as shown in the illustration. The seam on the branch pipe is made on the shortest side E,R'' in accordance with the usual practice. The seam on the main pipe may be taken at almost any point according to choice. In the illustration it is taken at C,G, on the opposite side to the branch pipe.

If it should be desired, as a matter of draughtsmanship, to determine the shape of the joint line between the two pipes in the elevation, this may be done by projecting back from the second projection through the first projection into the elevation. Thus, project the points on the semicircle E,F, through R'',S'',T'', to the ground line M'',N'',O''. Further, these points should be projected on to or located on the circle around S' in the first projection. Then, from the points on the circle, project lines into the elevation. The next step is to take the perpendicular heights from the ground line M'',N'',O'' to the points on the joint line R'',S'',T'', and mark them in the elevation up the corresponding lines from the ground line M,N,O. For example, the height c'',d'' in the second projection may be traced back to the point c',d' in the first projection and on to the elevation to be located at the height c,d in that view. The top extremity d is then one point on the contour of the joint line. This process repeated with all the other perpendicular heights from the second projection should give sufficient points in the elevation through which to draw the contour of the joint line as shown in the illustration at Fig. 217.

AN OFF-CENTRE PIPE ON OBLIQUE CONE

While a cylindrical pipe intersecting an oblique cone in an off-centre position is by no means a difficult problem when arranged in a convenient position for development, it may prove very perplexing when presented on a working drawing to suit given conditions. In the illustration at Fig. 218, the elevation and plan show an oblique conical reducing piece intersected by a cylindrical pipe. The true angle or inclination at which the pipe meets the cone is not seen in either the plan or the elevation, but in the first projection the true

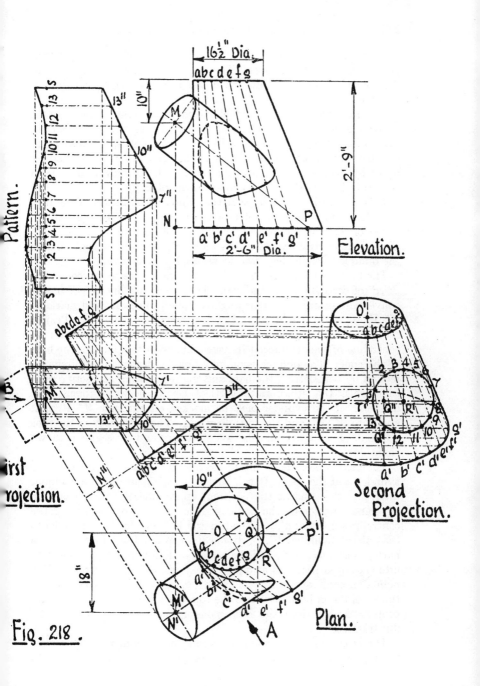

Pattern.

$16\frac{1}{2}''$ Dia.

a b c d e f g

M

$10''$

$13''$

$10''$

$7''$

N

$2'-9''$

a' b' c' d' e' f' g'
$2'-6''$ Dia.

P

Elevation.

S
13 12 11 10 9 8 7 6 5 4 3 2 1
S

O''
a b c d e f g

2 3 4 5 6
7

M'
7'
13'
10'
P''
a b c d e' f' g'

T''
Q''
R''
8

14
15
9
13 12 11 10'
a' b' c' d' e' f' g'

First
Projection.

N'''

B

Second
Projection.

$19''$

T
O
Q
P'
a b c d e f g
R

$18''$

M''
N''
a'
b'
c'
d' e' f' g'

A

Plan.

Fig. 218.

angle is obtained between the plane of the base of the cone and the central axis of the cylinder. The pattern for the cylinder may be "unrolled" from this view when the shape of the joint line is determined.

To obtain the true angle of inclination of the cylinder, produce the centreline M,P of the cylindrical pipe in the elevation to meet the plane of the base at P. The position of this point, which may occur inside or outside of the actual base of the cone, is located in the plan at P' by dropping a vertical line from P in the elevation to meet the centreline $M'P'$ of the cylinder in the plan.

It is assumed, of course, that the ordinary conditions of the plan and elevation have been set down in accordance with the requirements of the working drawing, and therefore the position and direction of the central axis of the cylindrical pipe and the centreline of the oblique cone in these two views are taken as given.

The first projection is now taken at right angles to the centreline M',P' of the cylinder in the plan. Draw the base line N'',P'' in any convenient position at right angles to the projectors, and mark off N'',M'' equal in height to N,M taken from the elevation. Project P' to the base line, to obtain P''. The line $M''P''$ then represents the centreline of the cylinder at its true angle with the plane of the base of the cone. The oblique cone may now be drawn in the first projection by transferring the base and the top to their respective positions with the appropriate vertical height between them. The diameter of the cylinder may be marked off through M'' at right angles to M'',P'', and the pipe drawn in. The extremity of the pipe above M'' bisects the angle between the pipe and the segment shown chain-dotted.

The next process is to obtain the line of intersection between the cylinder and the cone. It will be seen in the plan that the cylinder intersects the cone at the position, approximately, of the bottom left-hand quarter of the conic surface. Therefore, regarding the centrelines a,O and a',Q of the top and bottom circles as parallel to that of the cylinder M',P', divide the top quadrant into six equal parts, as at a,b,c,d,e,f,g and the corresponding bottom quadrant also into six equal parts, as at a',b',c',d',e',f',g'. Join the corresponding points on these quadrants, as at a,a'; b,b'; . . . g,g'. Now transfer these points on both circles to the top and bottom edges in the first projection, and join the corresponding points as shown in the illustration.

The next step is to obtain the second projection, which is a view looking straight into the branch pipe in the direction of arrow B.

Produce the centreline M'',P'', and mark off the centre-point R' in any convenient position. The point R' is the centre of the circle representing the branch pipe. Now project the circular top and bottom of the cone from the first into the second projection, to obtain the respective ellipses around the centres O' and Q'. An important point should be observed here. The off-centre distances between the three centres, R',Q', and O', may be obtained from the plan. From the point R on the centreline of the branch pipe in the plan, the perpendicular distances of the bottom and top centres are R,Q and R,T, respectively. Take these distances and mark them off in the second projection back from R' on the projected centreline, as shown at R',Q'',T'. Thus, the centre Q' lies on the perpendicular below Q'', and the centre O' lies on the perpendicular above T'.

Assuming, now, that the six points have been obtained on the top and bottom ellipses, and joined as shown in the diagram, it will be seen that the line joining the points on the ellipses cross the circle around R'. The points where these lines cross the circle may now be projected back into the first projection to meet the corresponding lines on the conic surface in that view. For example, where the line d,d' crosses the circle at point 10, this point is projected back into the first projection to meet the line h,h', in point 10' in that view. Similarly, where the line g,g' cuts the circle at point 7, and the line a,a' crosses at point 13, projectors back to the corresponding lines g,g' and a,a' in the first projection will give points 7' and 13'. This process repeated with all the other points around the circle in the second projection will afford sufficient through which to draw the joint curve in the first projection. Preparatory to the next operation, all the projector lines from the circle to the joint curve should be produced to meet the other end of the pipe beyond point M''.

The pattern may now be "unrolled" from the branch pipe in the first projection. Project the diameter or girth line through point M'' into the pattern, and, beginning at any convenient spot, mark off the distances, $s,1,2,3,\ldots 12,13,s$, equal to those around the circle in the second projection. It will be observed that the seam is taken at the position of point s on the horizontal centreline of the circle. The spacings thus marked off along the girth line in the pattern are all different in length, according to the various positions at which the conic lines across the circle in the second projection, but when all marked off in their correct order the total length of the girth line should be equal to the circumference of the circle. Next, from all the points on the joint curve, and also all the points on the other end of the pipe beyond M'' in the first projection, project lines into the pattern parallel to the girth line. Through all the points on the girth line draw lines at right angles to it to meet the corresponding lines parallel to it, when curves drawn through the appropriate points of intersection will give the contour of the pattern, as shown in the illustration.

The joint curve in the plan and elevation may be obtained from the first projection by locating the points on the conic lines a' to g' in their respective positions in each view. For instance, the vertical height of each point from the plane of the base will be the same both in the elevation and the first projection. It is, therefore, only a matter of transferring these heights from one view to the other in order to obtain the required points on the corresponding conic lines through which to draw the curve.

CHAPTER 16

TWISTED SURFACES

PROBLEMS of twisted surfaces are generally more perplexing than
difficult. Some surfaces appear to be twisted when, in fact, they are
merely rolled. In the projected view of the transition piece shown
in Fig. 221 the surfaces have every appearance of being twisted, but
a careful study of the front and side elevations will show that each
of the four sides of which the transition piece is composed is bent or
rolled to a portion of a cylindrical surface. The example shown in
Fig. 220 is a problem of a similar character, but presents an appearance more akin to a swan-neck. Nevertheless, the four sides com-
posing the transition piece need only bending or rolling in order to
bring them to the required shape.

In these two examples it is important to observe that while the
duct as a whole constitutes a twist, the plates which form the sides

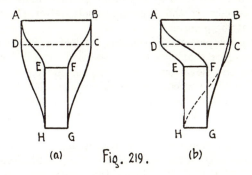

(a)   Fig. 219.   (b)

of the duct are not themselves twisted. The difference between an
apparent twist and a real twist is shown in a somewhat simple illustration in Fig. 219. For instance, in the example at (a), the rectangle
A,B,C,D is turned through 90 degrees to the position E,F,G,H, a
similar rectangle at the other end, but the plates forming the sides
of the duct are clearly not twisted. This may be seen best in the top
plate A,B,F,E, which joins the two edges A,B and E,F. Now, in
the example at (b), the two rectangles A,B,C,D and E,F,G,H occupy
the same positions relative to each other as do those in the example
at (a), but it will be seen that the plate from the edge A,B is really

twisted to join the corresponding long side of the other rectangle at F,G. The example at (b) may be regarded as similar to a twisted rectangular bar, in which the four faces of the bar must actually twist to the new position. The difference in the conditions exemplified by these two examples should be clearly grasped, as the successful solution of a problem often depends on the satisfactory differentiation between a real and apparent twist.

SWAN-NECK TRANSITION PIECE BETWEEN PARALLEL PLANES

Since the four sides or plates of the transition piece shown in Fig. 220 are rolled surfaces, the patterns may be developed by the Parallel Line Method. The four surfaces are lettered A,B,C,D in clockwise rotation. In the right-hand elevation the front surface is marked A, the right-hand side B, the back surface is marked with the dotted letter C, and the left-hand side with D. The vertical height between the two flanges is divided into six equal parts, and numbered as at 1,2,3,4,5,6,7. Through these points, horizontal lines are drawn across both elevations, which may represent parallel lines drawn right round the surface of the transition piece, after the manner of horizontal cutting planes.

To develop the pattern for the side marked A, in the right-hand elevation, imagine that side to be stretched or rolled out flat, as shown above the elevation. The width of the pattern along the parallel lines will be the same as seen in the elevation, but the height vertically upwards will be equal to the length of the outside contour shown at A', in the left-hand elevation. Therefore, take the distances between the points $1',2'$; $2',3'$; $3',4'$; $4',5'$; $5',6'$; $6',7'$ and mark these off in their respective order along the vertical line produced from the edge of the rectangle in the right-hand elevation, as seen at $1'',2'',3'',4'',5'',6'',7''$. From the points marked off, draw parallel lines horizontally across to meet another set drawn vertically upwards from the parallel lines which cut the sides in the elevation. The contour of the pattern may now be plotted through the points where the lines meet, as shown in the diagram.

The pattern for side C, shown below the right-hand elevation at C'', is obtained in a similar manner. In this case the spacings along the contour at C' in the left-hand elevation are marked off along the vertical line produced downwards below the edge of the rectangle in the right-hand elevation. Horizontal lines drawn across to meet the lines dropped vertically downwards from the elevation will

produce the points through which the contour of the pattern may
be drawn.

The pattern for the side D is shown above the left-hand elevation
at D'', and is developed in the same way by spacing the distances

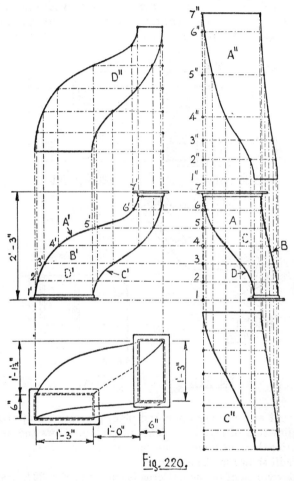

Fig. 220.

vertically upwards, this time equal to those along the contour of
the curve at D in the right-hand elevation. The procedure is otherwise the same as for the other two patterns. The pattern for the
remaining side, B, is not shown, but no difficulty should be experienced if the same routine is followed, using the spacings from the

contour of the curve at B in the right-hand elevation in order to obtain the true height of the pattern.

TRANSITION PIECE BETWEEN PLANES AT RIGHT ANGLES

The transition piece shown in Fig. 221 is somewhat similar to that of the previous problem except that it joins two rectangles in planes at right angles to each other. The illustration at Fig. 221, in addition

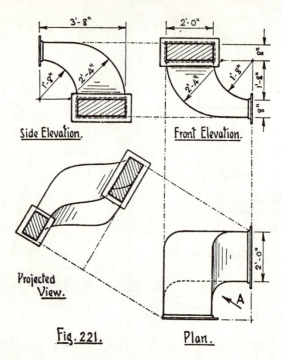

Side Elevation.

Front Elevation.

Projected View.

Fig. 221.

Plan.

to the front and side elevation and plan, presents a projected view in the direction of arrow A. This projection shows the apparent twist better than the other views, but the surfaces are actually rolled to the radii shown in the elevations.

The development of the patterns is shown in Fig. 222. As in the previous example, a series of horizontal lines is drawn through both elevations, as at 1,2,3,4,5,6, to represent contour lines right round the transition piece. The four surfaces are lettered A,B,C,D. In the right-hand elevation the front surface is marked A, and in clockwise rotation the right-hand surface becomes B, the back surface is marked with the dotted letter C, and the left-hand surface with D.

Now, taking the front surface A, imagine it to be flattened out as seen above at A''. The lengths of the horizontal parallel lines across

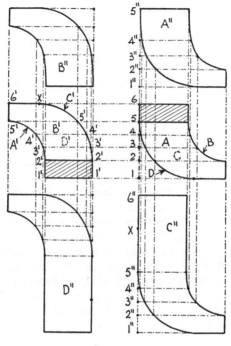

Fig. 222.

the pattern will be the same as seen in the elevation below, but the height of the pattern, from point $1''$ to $5''$, will be equal to the length of the curve at A' in the left-hand elevation. Therefore, from the curve A' take the distances from $1'$ to $5'$, and mark them off up the vertical line from the edge of the front side A in the right-hand elevation, as at $1'',2'',3'',4'',5''$. From these points draw parallel lines horizontally across to meet the vertical lines drawn from the points on the curves in the elevation below. Thus, points will be obtained through which to draw the contour of the pattern as shown at A'' in the diagram.

The pattern C'' below the elevation is for the back side C. The distances spaced off down the vertical line from the edge of the elevation are taken from the contour, marked C', in the left-hand elevation. It will be seen that an extra point, X, is placed between $5'$ and $6'$. This is to ensure greater accuracy in the spacing between those two points. Otherwise the method of drafting the pattern is precisely the same as for the front side A.

The patterns for the sides B and D are obtained from the left-hand elevation, and are shown above and below, at B'' and D''. The vertical spacings are taken this time from the right-hand elevation, and are equal to the distances round the curves B and D respectively.

ELBOW TRANSITION PIECE BETWEEN ANGULAR PLANES

When a connecting duct is required between two rectangular frames which occupy relative positions such as those shown at A,B,C,D and E,F,G,H (Fig. 223) the geometry of the construction might take numerous forms. The design here given is one which takes a smooth bend, thus offering a minimum of resistance to air flow, and is, moreover, of pleasing appearance when it is made. The illustration at Fig. 223 shows the plan and elevation as they would be given on a working drawing, and also a projected view which presents the full radii of the inner and outer curves.

It may be noticed at this point that the bend is made up of eight sides instead of four. For example, in the elevation, perhaps the most obvious is the triangular side with its base at F,G and its apex at A. The two triangular faces on either side of this are, first, the one with its base at A,B and its apex at F, and, second, that with its base at A,D and its apex at G. These two triangular sides are reversed in position in respect to base and apex. The next two triangles on either side of these are represented by mere straight lines in the elevation. Thus, the line EF,B at the top and GH,D at the bottom represent the triangular faces, which are best seen in the projection, at E,F,B and G,H,D. Moreover, it will be seen that in the projection the surface G,H,D lies below, or is exactly covered by, the surface E,F,B. In the elevation, the next two triangles on either side of these are, first, the one with B,C as its base and its apex at E and, second, that with its base at C,D and its apex at H. Finally, the remaining triangle has its base at E,H and its apex at C.

It is important to locate these triangles carefully before going on to the pattern-development, as the successful solution depends on a clear vision of these eight faces. It will be noticed that each face has one straight side and two curved ones, in triangular form.

In preparation for developing the pattern, a series of parallel lines is drawn round the bend, as shown in Fig. 224. This case is not quite so simple as the two previous examples, and will need a little care in following the lines round the body of the bend in the elevation. First, divide the outer curve in the projected view into a number of equal parts, as at F,a,m,A. In the illustration it is divided

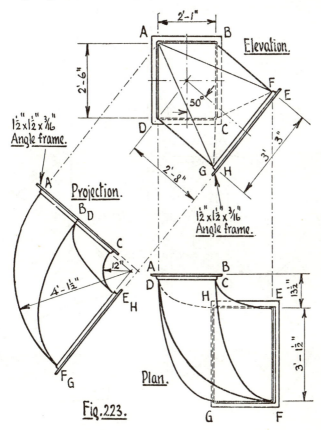

Fig. 223.

into three only, in order to avoid undue congestion of lines. Project the points a and m back into the elevation to cross the triangle F,G,A, in points a,b and m,n. The short lines a,b and m,n obtained on this triangle are the starting points for the contours round the surface of the bend. For example, from each extremity of the line a,b draw lines parallel to the two bases A,B and A,D to meet the sides of the next two triangles in *c* and *h*. The next two lines in this contour in the elevation are really at right angles to the plane of the paper. It will be remembered that the outside line *EF,B* represents one of the faces, as seen full in the projected view at *E,F,B*. Thus, in the projected view, the line between the two points marked *c* and *d* is the line which, in the elevation, is at right angles to the plane of the paper, and is represented by the single point marked *cd*   This

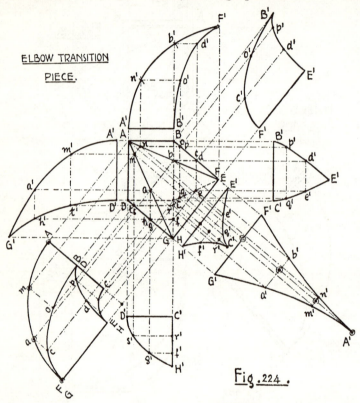

ELBOW TRANSITION
PIECE.

Fig. 224.

reasoning applies also to the bottom face at *GH,D*, in which the point marked *gh* represents a similar line at right angles to the plane of the paper. Now, from these two points, *cd* and *gh*, in the elevation, lines should be drawn parallel to the two bases *B,C* and *C,D* to meet the sides of the next two triangles in points *e* and *f*. Finally, the remaining line should be drawn between the two points *e* and *f*, parallel to the base line *F,G*.

The second contour, derived from point m, is plotted round the surface of the bend in precisely the same way as that just described in connection with point a, and by a careful inspection of the diagram the contour m,n,o,p,q,r,s,t,m in the elevation should be readily followed.

The patterns are developed by the parallel line method and may be obtained from the elevation. Considering, first, the triangle A,F,G, the direction of the projection for the pattern is taken at right angles to the base line F,G, and the perpendicular distance from the base line F',G' to the point A' in the pattern is equal to, and taken from, the curved back, F,a,m,A, in the projected view. Thus, the spacings F,a; a,m; m,A are taken and marked off along the line in the pattern projected from the point A. Through these points, which are marked with a ring round them to make them clear, draw lines parallel to the base line F',G'. From the other points a,b,m,n, on the triangle A,F,G, in the elevation, project lines into the pattern to cross the parallel lines and locate the points a',b',m',n'. Curves drawn through G',a',m',A' and F',b',n',A' will give the pattern for that side.

The next triangle to be developed is that at E,H,C in the elevation. This pattern is projected at right angles to the base line E,H, and the distances marked off along the line projected from C are this time taken from the inner curve E,C in the projected view. These spacings will be seen marked by heavy dots in the pattern from the line E',H' to the point C'. Through these dots draw lines parallel to the base line $E'H'$, and from the points e,f,q,r, on the side of the triangle in the elevation, project lines into the pattern to meet the parallel lines in e',f',q',r'. Curves drawn through points E',e',q',C' and H',f',r',C' will complete the pattern for that side.

The next pattern to receive attention will be that for the side A,B,F above the middle triangle. From all the points on the sides A,n,b,F and B,o,c,F, project lines at right angles to the base A,B into the pattern above. Mark off the pattern base line A',B' in any convenient position. The next step is an important one. It will be observed that the joint line between the middle triangle A,F,G and the top triangle A,B,F occurs along A,n,b,F. The length of this joint line must be the same on both patterns, otherwise they would not fit. This length is already obtained in the pattern for the middle triangle, as is represented by the curve A',n',b',F' in the pattern. Therefore, all that is now necessary is to take the distances between those points and mark them off in their respective positions in the pattern for the top triangle. For example, take the distance A',n' and, from point A' in the pattern at the top, describe an arc cutting the vertical line from n in point n'. Next take the distance n',b' from the first pattern, and from point n' in the top pattern describe an arc cutting the vertical line from b in point b'. Now take the remaining distance b',F' from the first pattern, and from point b' in the top pattern describe an arc cutting the vertical line from F in point F'. It may now be remembered that, in the elevation, the lines n,o and b,d are drawn parallel to A,B. Therefore, in the pattern above, draw lines from n' and b' parallel to A',B', to cut the vertical lines from o and d in points o' and d'. Curves may now be drawn through A',n',b',F' and B',o',d',F' to complete the pattern for the top triangle.

The next pattern may now be drafted for the side A,D,G on the left of the middle triangle. The process is similar to that described for the top triangle, in so far as lines are projected into the pattern from points A,m,a,G and D,t,h,G in the elevation. The length of the curved side A',m',a',G' may then be obtained by taking the distances A',m'; m',a'; a',G' direct from the first pattern, and stepping these over in their respective order in the left-hand pattern. The reason for this is the same as in the previous case, that is, these two curved edges in the respective patterns must fit together along the joint line A,m,a,G when the sides are formed and fitted together. The lines m',t' and a',h' in the left-hand pattern are drawn parallel to A',D'. Curves drawn through the points A',m',a',G' and D',t',h',G' will then complete the pattern.

The patterns for the next two sides EF,B and GH,D, which lie in planes at right angles to that of the paper, may be obtained direct from the projected view. Thus, the side F,c,o,B,p,d,E, seen in full face, is the true pattern required for both sides. This outline is shown repeated as a pattern above the elevation at $F',c',o',{}'B,p',d',E'$.

The two remaining patterns B',p',d',E',e',q',C' and C',r',f',H',g',s',D are developed in precisely the same manner as those obtained at the top and on the left of the elevation, and should be readily followed from the diagram.

SPIRAL FINIAL

The problems of apparently twisted surfaces so far considered have been developed by the parallel line method. Those which follow are dealt with by triangulation.

The spiral finial shown in Fig. 225 is twelve-sided and the amount of twist seen in the elevation is governed by the radius of the curves in the plan. The outer circle in the plan is divided into twelve equal parts, and the oval segments are drawn, one on each centreline. The vertical centreline in the elevation is divided into any number of equal parts, and horizontal lines drawn cross to the outside curves. These lines may be regarded as horizontal contours round the surface of the finial, and may be located in the plan by dropping vertical

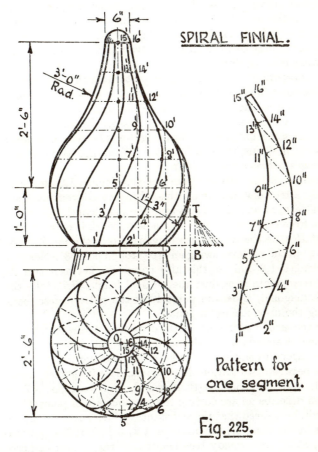

SPIRAL FINIAL.

Pattern for one segment.

Fig. 225.

lines from the points on the outside curve in the elevation to the horizontal centreline in the plan. Then, with O as centre, circles drawn from these points will represent the contour lines in the plan.

The spiral shapes of the joint lines in the elevation may now be obtained from the plan. Where the contour circles in the plan cross the oval joint lines, the points may be projected vertically upwards to the corresponding contour lines in the elevation. For example, in the plan, the oval on the vertical centreline, $O5$, is crossed by the contour lines in points 1,3,5,7,9,11,13,15. Then these points projected vertically upwards to the corresponding contour lines in the elevation will give the points 1′,3′,5′,7′,9′,11′,13′,15′, through which the spiral curve may be drawn. This process repeated with the points on the next oval, on the centreline $O,6$, should produce the points 2′,4′,6′,8′,10′,12′,14′,16′, in the elevation, through which that spiral curve may be drawn. In this way all the spiral curves in the elevation may be drawn.

In preparation for developing the pattern for the segment between the two joint lines 1 to 15 and 2 to 16, the plan of this segment is triangulated as shown by the zigzag line 1,2,3,4, . . . 13,14,15,16. In the elevation, since the vertical distance between each contour line is the same, only one vertical height will be needed in developing the pattern. Thus, the first height B,T from the base is used for all the stages between the contours.

To develop the pattern, take the plan length 1,2 and, as this is already a true distance, mark this off in any convenient position in the pattern, as at 1″,2″. Next take the plan length 2,3 and, as this passes from one contour line to the next, mark it off along the base line at B, take the true length diagonal up to the top T, and, from point 2″ in the pattern, swing an arc through point 3″. Next take the plan length 1,3 and, as this also passes up between two contour lines, triangulate it against the vertical height B,T, take the true length diagonal, and from point 1″ in the pattern describe an arc cutting the previous arc in point 3″.

For the second triangle, take the distance 3,4, direct from the plan, and, as it is already a true length, from point 3″ in the pattern swing an arc through point 4″. Next take the plan length 2,4 and, as this passes between two contour lines, triangulate it against the vertical height, take the true length diagonal, and from point 2″ in the pattern describe an arc cutting the previous arc in point 4″.

It will be noticed that those distances which occur on the contour circles in the plan, as, for instance, 1,2 ; 3,4 ; 5,6 ; 7,8 ; 9,10 ; 11,12 ; 13,14 ; 15,16, are already true lengths, as they occur in horizontal planes and therefore have no vertical heights. These distances are taken direct for marking off in the pattern. All the other plan lengths which pass between two contour circles must be triangulated against the vertical height to obtain the true lengths for the pattern.

The remainder of the pattern is straightforward triangulation, and by following the rules of that method should not be difficult to complete.

LOBSTER-BACK SPIRAL

The problem represented in Fig. 226 is of a spiral lobster-back, and is typical of many examples where pipes have to form a bend round a central column, or take a curve round a corner at an inclined

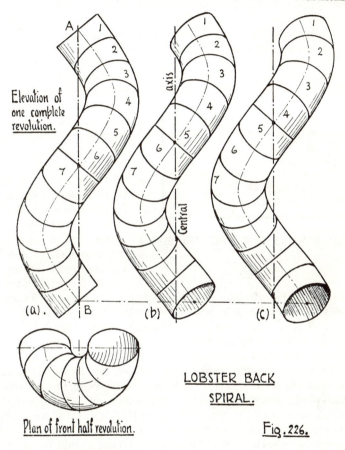

Elevation of one complete revolution.

(a). B (b) (c)

LOBSTER BACK
SPIRAL.

Plan of front half revolution. Fig. 226.

angle. The spiral lobster-back may also be used as a chute for gently conveying parcels or material from a point above to another vertically below, thus avoiding a direct fall.

Before attempting the pattern-development, it is important to grasp one or two important principles governing the construction of this type of chute. The pitch is the distance between the points of one complete revolution, as from *A* to *B*, shown in Fig. 226 (*a*).

The angle of inclination of the chute is obtained by placing the circumference at right angles to the pitch, as illustrated at Fig. 227. It is important to observe that when this chute is revolved on the central axis of the spiral, each joint line between the segments, as it comes round to the front position, presents a straight line which is always at right angles to the line of inclination. This may be seen by an inspection of the three figures at (a), (b), (c), Fig. 226. In this example there are twelve segments to one complete revolution. In the figures at (b) and (c) the spiral is revolved through one-twelfth of a revolution, thus bringing the straight joint line between the segments one joint higher up each time. In the figure at (a) this joint occurs between segments 6 and 7. In the figure at (b) the straight joint line has moved up to that between segments 5 and 6. In the figure at (c) a further one-twelfth revolution brings the straight joint line between the segments 4 and 5.

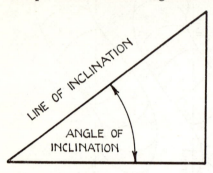

CIRCUMFERENCE OF PLAN CENTRELINE.

Fig. 227.

Moreover, the joint line is always at the same angle. A full appreciation of this principle is essential before considering the problem of developing the pattern.

The drafting of the pattern is given in Fig. 228. One segment only, that above the straight joint line, is set out for the development. First draw the semicircle A,G in the plan to represent the centreline of the spiral; also draw the inner and outer semicircles to suit the diameter or size of the pipe. Since, in this case, there are twelve segments in one complete revolution of the spiral, the semicircle A,G, which represents one half of a revolution, should be divided into six equal parts, as will be seen lettered A,B,C,D,E,F,G in the diagram.

The next step is to determine the angle and position of the straight joint line $7',19'$ in the elevation. To do this, set out the base line A',G', and on either side of the vertical centreline mark off three divisions equal to those round the semicircle. Thus the length of A',G' should be equal to the length of the semicircle A,G. From the point G' erect a vertical line, and mark off the height G',P, equal to half of the pitch of the spiral. Join A',P. Then the line A',P represents the inclination of the chute, and P,A',G' the angle of inclina-

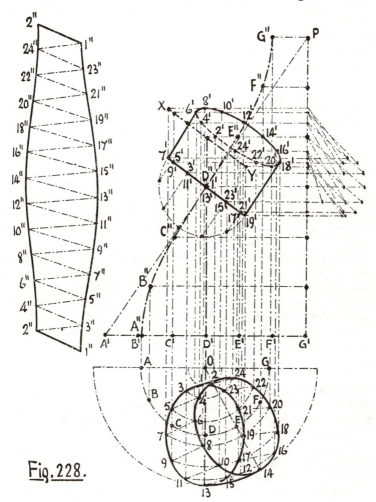

Fig. 228.

tion. The straight joint line $7',19'$ may now be drawn at right angles to A',P through the point where the vertical centreline crosses the line A',P.

Now, since there are to be six segments in the semicircle in the plan, there must be six segments in the half-spiral in the elevation. Therefore, divide the half-pitch G',P into six equal parts, and from the points obtained thereon, draw horizontal lines to meet a set of vertical lines drawn from the points A,B,C,D,E,F,G in the plan. The points where these two sets of lines meet should enable the spiral curve $A'',B'',C'',D'',E'',F'',G''$ to be drawn in. The point D'' occupies the centre of the joint line $7',19'$. Nevertheless, D'' is not actually on the joint line, but represents the centre-point of the circle, or is on the central axis of the segment around which the joint line $7',19'$ is formed. The other edge, or joint line, of this segment is represented by the ellipse $6',18'$ and it will be observed that the next point E'' on the spiral is the centre-point of the ellipse.

To obtain the plan of the segment, on the joint line $7',19'$, describe a semicircle, divide it into six equal parts and project the points perpendicularly back to $7',19'$. This joint line may now be dropped into the plan to obtain the ellipse $7,19$. It is important now to see that, as the joint lines between the segments occur in their respective positions round the spiral, the elliptical form of each will be exactly the same in the plan. Therefore, to obtain the upper edge, or joint line $6',18$, of this segment, all that is necessary in the plan is to reproduce the first ellipse on a centreline through E, at 30 degrees from the vertical centreline through D. This may be done by swinging arcs, using O as centre, from all the points on the first ellipse, and, on either side of the new centreline O,E, cutting them off equal to the respective arcs on either side of the vertical centreline O,D.

The next step is to obtain the ellipse $6',18'$ in the elevation which represents the top edge of the segment. It may be recalled now, from the consideration of the illustration in Fig. 226, that if the joint line $6',18'$ be turned back through one-twelfth of a rotation, it would appear as the straight line X,Y, vertically above $7',19'$, and identical with it in the matter of length and the points on it. Therefore, to produce the ellipse $6',18'$, transfer the line $7',19'$ and all the points on it, vertically upwards to the position X,Y. The vertical distance $7',X$ is equal to the vertical pitch of one segment; that is, equal to one of the divisions up the line G',P. Now, from the points on X,Y, draw horizontal lines to meet a set of corresponding vertical lines from the points on the ellipse $6,18$ in the plan. For example, to follow this process on one set of points, consider the point 3 on the first ellipse in the plan. The corresponding point on the second ellipse is point 4. The vertical line from point 3 passes up to point $3'$ on the line $7',19'$, and further upwards to meet the line X,Y. From the line X,Y a short horizontal line meets the vertical line from point 4 in point 4'. This process repeated with all the points on the ellipses in the plan will result in the second ellipse being obtained in the elevation as the ellipse 6',18'.

To develop the pattern, which is obtained by triangulation, divide the surface in the plan into triangles, as shown numbered at 1,2,3,4, . . . 21,22,23,24. For the sake of clearness, number the corresponding points in the elevation as shown at 1',2',3',4', . . . 21',22',23',24'. All these points in the elevation should now be projected horizontally to a vertical height line, as, in the illustration, to the vertical line G',P. The points on the line 7',19' may be produced beyond the vertical height line to serve as base lines along which to mark the plan lengths.

For the first triangle, take the plan length 1,2, and mark it off along the base line level with point 1'. Take the true length diagonal up to the point level with 2', and in any convenient position to begin the pattern, mark off 1",2". Next take the plan length 2,3, and mark it off along the base line level with point 3'. Take the true length diagonal up to the point level with 2', and from point 2" in the pattern swing an arc through point 3". The next distance in the pattern 1",3" required to complete the first triangle should be taken from the semicircle on the line 7',19'. These are the true spacings round the circular edge of the segment.

For the second triangle, take the plan length 3,4 and mark it off along the base line level with point 3'. Take the true length diagonal up to the point level with point 4', and from point 3" in the pattern swing an arc through point 4". To complete the triangle, again take one of the spacings from the semicircle and from point 2" in the pattern describe an arc, cutting the previous arc in point 4".

For the third triangle, take the plan length 4,5, and mark it off along the base line level with point 5'. Take the true length diagonal up to the point level with point 4', and from point 4" in the pattern swing an arc through point 5". Again take one of the true spacings from the semicircle, and from point 3" in the pattern describe an arc cutting the previous arc in point 5".

The remainder of the pattern is straightforward triangulation and, with care in using the correct vertical heights, should be easily completed by following this routine. It should be observed that all the spacings along both sides of the pattern are equal, and taken from the semicircle on the line 7',19'. All the diagonals are triangulated against their appropriate vertical heights.

This type of chute is simple and easy to make, as no distortion of the segments, other than rolling, is necessary for fitting together.

SPIRAL BLADES

The screw conveyer blade, generally made of 10, 12, or 14 gauge mild steel, is an example of a surface really twisted. The illustration at Fig. 229, shows a spiral blade of this kind. These are usually pressed hot over a forming die, but there are often times when a blade or surface such as this has to be made by hand. It is then that a clear understanding of the properties of this spiral form is essential if the work is to be executed quickly and efficiently.

A true geometrical pattern for this blade cannot be developed, but

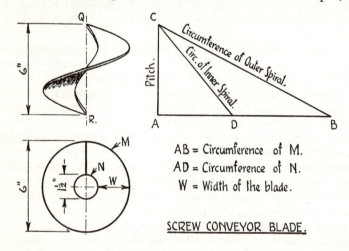

AB = Circumference of M.
AD = Circumference of N.
W = Width of the blade.

SCREW CONVEYOR BLADE.

Fig. 229.

an approximation can be made, which, when worked correctly, will reach the form required.

Imagine the blade shown in the elevation, Fig. 229, to be pressed down flat. The result would be a disc with a hole in the centre. This process is, of course, the reverse of that required to form the blade, but it serves to show that the pattern should be a circular blank with a centre hole, similar to the illustration in Fig. 231. The length of the outer circumference of the blank should be equal to the length of the outer edge of the spiral. The length of the outer edge of the spiral may be found by calculating the length of the hypotenuse of a right-angled triangle, as shown at A,B,C, Fig. 229, where A,B represents the circumference of the plan circle M and A,C equals the pitch of the spiral. Then the hypotenuse BC represents the length of the outer spiral from Q to R.

The length of the inner edge of the spiral may be found in the same way, by putting A,D equal to the circumference of the centre hole N in the plan, and A,C the pitch of the spiral. In this case the hypotenuse D,C represents the length of the inner spiral from Q to R.

This principle may be verified by cutting a right-angled triangle

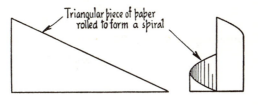

Triangular piece of paper
rolled to form a spiral

Fig. 230.

from a piece of paper, and rolling it on to a cylinder of appropriate diameter, as illustrated in Fig. 230.

Now, if the length of B,C be turned into the circumference of a circle, the diameter should give the size of the disc required for the pattern. Again, if the width of the blade be marked off on the inside of the circle, and an inner circle drawn to suit this width, as shown in Fig. 231, then the circumference of the inner circle should correspond to the length D,C of the inner spiral. But it will be found that

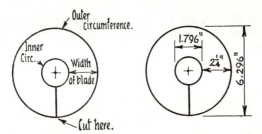

Outer Circumference.

Inner Circ.

Width of blade

Cut here.

1.796"

$2\frac{1}{4}$"

6.296"

Fig. 231.

the circumference of the inner circle falls considerably short of the length DC. In practice, therefore, there is no alternative but to stretch, or deform the metal in order to obtain the required shape.

A worked example will probably illustrate this point more clearly.

EXAMPLE. The outside diameter of a spiral is 6 in.; the centre diameter, or size of the shaft around which the spiral fits, is $1\frac{1}{2}$ in.; and the pitch of the spiral is 6 in. Calculate the size of the blank and compare the inner circumference with the inner edge of the spiral

$$\text{Here, the pitch} \quad AC = 6 \text{ in.}$$

$$\text{Circumference of} \quad M = \pi d$$

$$= 3 \cdot 1416 \times 6$$

$$= 18 \cdot 8496 \text{ in.}$$

$$\therefore \text{ Outer edge of spiral,} \quad BC = \sqrt{6^2 + 18 \cdot 8496^2}$$

$$= \sqrt{36 + 355 \cdot 3}$$

$$= \sqrt{391 \cdot 3}$$

$$= 19 \cdot 78 \text{ in.}$$

$$\text{Again,} \quad AD = \pi d$$

$$= 3 \cdot 1416 \times 1\tfrac{1}{2}$$

$$= 4 \cdot 712 \text{ in.}$$

$$\therefore \text{ Inner edge of spiral,} \quad DC = \sqrt{6^2 + 4 \cdot 712^2}$$

$$= \sqrt{36 + 22 \cdot 2}$$

$$= \sqrt{58 \cdot 2}$$

$$= 7 \cdot 629 \text{ in.}$$

If, now, a disc be cut for the pattern, having a circumference equal to that of the outer edge of the spiral, BC, the diameter of that disc will

$$= \frac{BC}{\pi}$$

$$= \frac{19 \cdot 78}{3 \cdot 1416}$$

$$= 6 \cdot 296 \text{ in.}$$

Since the diameter of the spiral is 6 in. and the centre hole $1\tfrac{1}{2}$ in., the width of the blade must be $2\tfrac{1}{4}$ in., or $4\tfrac{1}{2}$ in for both sides. Deducting this from 6·296 we have

$$6 \cdot 296 - 4 \cdot 5 = 1 \cdot 796 \text{ in.}$$

which is the diameter of the hole in the disc.

Here, then, comes the difficulty. The circumference of the 6·296 in. circle will conform to the length of the outer spiral, or 19·78 in., but the circumference of the 1·796 in. inner circle will *not* conform to the length of the inner edge of the spiral. The circumference of the 1·796 inner circle

$$= 1·796 \times 3·1416$$
$$= 5·642 \text{ in.}$$

But the inner edge of the spiral *DC* is shown to be 7·629 in. Therefore, an elongation of nearly 2 in. must take place on the inner edge

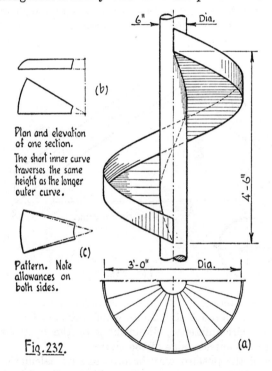

6″ Dia.

(b)

Plan and elevation of one section.

The short inner curve traverses the same height as the longer outer curve.

4′-6″

(c)

Pattern. Note allowances on both sides.

3′-0″ Dia.

Fig. 232.

(a)

of the blank in order to produce the correct spiral. This, of course, can be effectively produced by a hot pressing, but to work such a spiral cold, in sheet metal, causes much trouble and difficulty.

FLAT-BOTTOMED SPIRAL CHUTE

The type of spiral chute illustrated in Fig. 232 is one which is often used for parcels or sacks of grain or other material. The bottom of the chute is flat, or horizontal at any position on a line passing through the centre of the centre column. Geometrically, the bottom of this chute is exactly the same as that of the spiral conveyer blade. Constructionally, the method of making it differs according to the size and form of the blank. Generally, a chute of this kind varies from 12 in. to 2 ft. in the width of the bottom, and from 2 ft. 6 in. to 5 ft. or 6 ft. in overall diameter. For a chute of this size it is not practicable to make the bottom from full circular blanks. It is much easier to make it in short sections, as shown in Fig. 232.

The problem of stretching or beating the metal to make it conform to the required shape is the same for the large chute made in sections as it is for the smaller conveyer blade made from one circular disc.

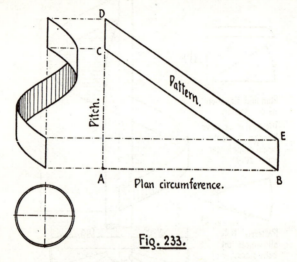

Fig. 233.

However, the fractional sections of the larger chute somewhat ease the difficulty of so much stretching, which needs to be a maximum at the inner circumference, diminishing as the operation moves outwards to the outside edge of the chute.

The size of the pattern may be obtained by calculation similar to the example given for the conveyer blade. The full circle may then be divided into sectors small enough to suit requirements. The illustration at (b), Fig. 232, shows a plan and elevation of one section only. It will be noted that in the elevation the section presents two parallel lines at the top and bottom. This shows that the short inner curve, which fits against the centre column, must traverse the same vertical height as the longer outer curve. The necessity for hammering and stretching on the inner side of the pattern should therefore be evident. With an allowance on each side of the pattern similar

to that shown at (c), Fig. 232, the amount of beating and deformation
may be reduced, and the fitting together of the sections somewhat
simplified.

The vertical side of this chute, which forms the outer rim, is a
very simple problem of development, and is equally straightforward
in the matter of construction. The illustration at Fig. 233 shows
the pattern for this form of spiral developed. The left-hand view
is an elevation of the vertical rim only, without the central column
and the flat bottom. The pattern on the right-hand side is obtained
by setting out AB equal to the length of the plan circumference,
and AC equal to the pitch of the spiral. The heights CD and BE
are each equal to the *vertical* height of the rim. Then the pattern
$BCDE$ only needs rolling on an axis parallel to the two ends CD
and BE, in order to produce the spiral form required for the rim of
the chute. The pattern for the full revolution, as shown in the figure,
may, of course, be made up of short pieces as desired.

INDEX